THE CASE OF THE MIDWIFE TOAD

On September 23, 1926, an Austrian experimental biologist named Dr. Paul Kammerer committed suicide. This formed the climax to a great evolutionary controversy which Kammerer's experiments had aroused. The battle was between the followers of Lamarck, who maintained that acquired characteristics could be inherited; and the neo-Darwinists, who upheld the theory of chance mutations preserved by natural selection. Dr. Kammerer's experiments with various amphibians, including salamanders and the midwife toad (*Alytes obstetricans*), lent much weight to the Lamarckian argument and drew upon him the full fury of the orthodox neo-Darwinists. Heading the attack on Kammerer was a British scientist, William Bateson, who hinted that his experiments were faked, but failed to examine the evidence, including the so-called nuptial pads of his last remaining specimen of the midwife toad. It was a young American scientist who delivered the coup-de-grâce. On a visit to Vienna, he discovered that the discolouration of the nuptial pads was due not to natural causes but to the injection of Indian ink. When his findings were published, Kammerer shot himself.

When Arthur Koestler, who has always been interested in this tragic story, decided to investigate the mystery, he expected to tell the tragedy of a man who had betrayed his calling; for Kammerer's suicide was accepted as a confession of guilt and his work was discredited from that day to this. Instead, as Koestler read the contemporary papers, corresponded with Kammerer's daughter, Bateson's son, and all the surviving scientists who attended Kammerer's lecture in Cambridge, he found himself writing a vindication of a man who was in all probability himself betrayed. The story that emerges is, on one level, a fascinating piece of scientific detection; and on another, a moving and human narrative about a much abused, brilliant and lovable figure. Though no Lamarckian himself, Koestler ends the book with an appeal to biologists to repeat Kammerer's experiments with an open mind in order to confirm or refute them. If Kammerer's claims were posthumously confirmed, our outlook on evolution would be significantly changed.

BOOKS BY ARTHUR KOESTLER

Novels
THE GLADIATORS*
DARKNESS AT NOON
ARRIVAL AND DEPARTURE*
THIEVES IN THE NIGHT*
THE AGE OF LONGING*

Autobiography
DIALOGUE WITH DEATH*
SCUM OF THE EARTH*
ARROW IN THE BLUE*
THE INVISIBLE WRITING*
THE GOD THAT FAILED (WITH OTHERS)

Essays
THE YOGI AND THE COMMISSAR*
INSIGHT AND OUTLOOK
PROMISE AND FULFILMENT
THE TRAIL OF THE DINOSAUR*
REFLECTIONS ON HANGING*
THE SLEEPWALKERS*
THE LOTUS AND THE ROBOT*
THE ACT OF CREATION*
THE GHOST IN THE MACHINE
DRINKERS OF INFINITY
SUICIDE OF A NATION (EDIT.)
BEYOND REDUCTIONISM: THE ALPBACH SYMPOSIUM
(EDIT. WITH J. R. SMYTHIES)

Theatre
TWILIGHT BAR

*Available in the Danube Edition

Paul Kammerer in 1924, age 44.

ARTHUR KOESTLER

The Case of the Midwife Toad

HUTCHINSON OF LONDON

HUTCHINSON & CO (*Publishers*) LTD
3 Fitzroy Square, London W1

London Melbourne Sydney Auckland
Wellington Johannesburg Cape Town
and agencies throughout the world

First published 1971

*This book has been set in Imprint type, printed in Great Britain
on antique wove paper by Anchor Press, and
bound by Wm. Brendon, both of Tiptree, Essex*

ISBN 0 09 108260 9

CONTENTS

ILLUSTRATIONS

ACKNOWLEDGEMENTS

I have the pleasant duty to thank all those who communicated to me their recollections of Paul Kammerer and the Vienna Institute: his daughter, Lacerta; Hans Przibram's daughters, Countess Vera Teleki and Mrs. Doris Baumann; his brother Professor Karl Przibram, University of Vienna; William Bateson's son, Professor Gregory Bateson, University of Hawaii; Professor Ludwig von Bertalanffy, State University of New York at Buffalo; Mrs. Bettina Ehrlich; Professor Karl von Frisch; Professor G. Evelyn Hutchinson, Yale University; Dr. L. Harrison Matthews, F.R.S.; The Hon. Ivor Montagu; Professor J. H. Quastel, F.R.S., University of British Columbia; Professor W. H. Thorpe, F.R.S., University of Cambridge, and Mrs. W. M. Thorpe; Professor Paul Weiss, the Rockefeller University, New York.

I am especially indebted to Professor Holger Hyden, University of Gothenburg, for carrying out an unorthodox experiment which shed new light on an old controversy.

I wish to thank the Librarian of the American Philosophical Society for his co-operation in making the Bateson Papers available; the Librarian of the German Institute, London, and Mr. Leon Drucker, bookseller, for tracing works by Kammerer; Mrs. J. St. G. Saunders, Writer's and Speaker's Research, Mrs. Evelyn Reynolds, The North-East London Polytechnic, and Mrs. Herta Howorka, Innsbruck, for their most efficient help.

The following authors and publishers have given their kind permission to quote passages from the following works: C. D. Darlington, *The Facts of Life* (George Allen & Unwin, 1953); the revised edition, published by Allen & Unwin in 1964, is entitled *Genetics and Man*. Sir Alister Hardy, *The Living Stream* (Collins, 1965). H. Graham Cannon, *Lamarck and Modern Genetics* (Manchester University Press, 1959).

One

In the early afternoon of September 23, 1926, a road-worker found the dead body of a well-dressed man in a dark suit on an Austrian mountain path. He was in a sitting position, back propped against a vertical rock face, right hand clutching the pistol with which he had shot himself through the head. One of his pockets contained a letter addressed 'to the person who finds my body'. It read:

> Dr. Paul Kammerer requests not to be transported to his home, in order to spare his family the sight. Simplest and cheapest would perhaps be utilisation in the dissecting room of one of the university institutes. I would actually prefer to render science at least this small service. Perhaps my worthy academic colleagues will discover in my brain a trace of the qualities they found absent from the manifestations of my mental activities while I was alive. Whatever happens to the corpse, buried, burned, or dissected, its bearer belonged to no religious community and wishes to be spared a religious ceremony, which probably would be denied him anyway. This is not meant to express hostility against the individual priest, who is human like everybody else and often a good and noble person.[1]

The letter was signed Dr. Paul Kammerer. A postscript requested his wife to abstain from wearing black clothes or other signs of mourning.

Thus ended the greatest scientific scandal of the first half of

13

our century. Its hero and victim was one of the most brilliant and unorthodox biologists of his time. He was forty-five years old when the joint pressures of an inhuman Establishment and his own all-too-human temperament drove him to suicide. He had been accused of the worst crime a scientist can commit: of having faked his experimental results. Yet an obituary article in *Nature*, which is probably the world's most respected scientific journal, called his last book 'one of the finest contributions to the theory of evolution which has appeared since Darwin'.[2]

2

I first became interested in Kammerer's ideas as a student in Vienna. I was twenty when he died, and had never met him, but through the years the thought of writing about him lingered in the back of my mind. The secret of his attraction lies partly in his complex character and tragic fate, but above all in his heretical ideas, which he tirelessly attempted to prove in his experimental work, papers in learned journals and popular books. Kammerer refused to accept the Darwinian theory of evolution based on random mutations—haphazard variations produced by blind chance; and believed that the main vehicle of progressive evolution was the 'inheritance of acquired characteristics' which Lamarck had postulated in 1809—that is to say, that useful adaptive changes in the parents were preserved by heredity and transmitted to their offspring. Now this was, and still is, an explosive issue among biologists; the controversy between Darwinians and Lamarckians has raged for nearly a century, charged with emotional, political, even theological passion, and conducted, as we shall see, with astonishing disregard for the rules of fair play. This was the intellectual climate which made the scandal blossom and end in tragedy.

Kammerer's undoing was a grotesque amphibian creature: the midwife toad *Alytes obstetricans*, or, more precisely, its so-called nuptial pads—small callosities with horny spines on the male's fore-limbs which give it a better grip on the female while mating. These pads Kammerer claimed to be proof of the inheritance of acquired characters, while his opponents denied their existence.

In the pages that follow we shall hear a lot more of the nuptial pads of the midwife toad. They provide all the makings of a thriller with a lurid end. It was actually turned into a film, in

Stalin's Russia, where the Establishment was committed by the Party line to the Lamarckian theory of evolution, in contrast to the Darwinism of the West. The film was called *Salamandra*; it was made immediately after Kammerer's death, and was so popular that I was still able to see it six years later when I was in Moscow. It had been written and directed by the People's Commissar for Public Education, Lunacharsky, whose wife played the heroine. The hero of the film was subjected to various ignominies by reactionary Darwinian scientists, aided for good measure by reactionary monks. This was an exaggeration: Kammerer was exposed only to academic venom. Vienna, as Freud could testify, had a justified reputation for it.

There is no biography of Paul Kammerer in existence. The sources for the pages which follow are his books and technical papers, the polemics to which they gave rise, documents in various archives, and personal communications from a number of people, scientists and others, who had known him—among them Kammerer's daughter, his only child, who was baptised by the name of Lacerta. The *Lacertidae* are a genus of pretty lizards which he was very fond of; he also discovered a previously unknown variety which is named after him—*Lacerta fiumana Kammereri*.

Lacerta Kammerer now lives in Australia, under her married name, and her letters convey an intimate portrait not only of her father but also of the life of the intellectual élite of Vienna before the First World War.

3

The obituaries in the Austrian Press provide a kind of bird's-eye view of Paul Kammerer's life, seen through contemporary eyes.

Neue Freie Presse, Vienna*—24.IX.1926.

News of a tragic event reached us late in the evening. The eminent biologist, Dr. Paul Kammerer, whose books and essays in biology and sociology attracted wide and justified attention, whose lectures were always attended by an enthusiastic audience of hundreds of people, has ended his life by his own hand. Dr. Kammerer shot himself on the

* The *Neue Freie Presse* was regarded as the Austrian equivalent of *The Times*—but was rather more ebullient in style.

Schneeberg. The letters he left provide no complete explanation of the reasons for his fatal decision.

Dr. Kammerer was twice married, both marriages ended in divorce. His first wife, daughter of the prominent politician and former member of the Reichsrat, Dr. von Wiedersperg, renounced for his sake a promising career on the stage; his second wife was a well-known, successful painter.

An Austrian Biologist Writes: One of the best-known biologists of Vienna has put a voluntary end to his life. A man whose importance derived not only from his wellnigh amazing knowledge of all branches of natural science, but also from his gift to express his knowledge in a manner accessible to the general public, to call attention to the gaps in our knowledge, and to point the way to new discoveries. Owing to this, his scientific career was unusually rich in success. The manner in which he made his discoveries was characteristic of his genius, grasping by intuition, as it were, the requirements for finding the answer to a general and even a specialised problem.

His development led from music to zoology and sociology. After matriculating from the *Gymnasium*, he entered the Vienna Academy of Music, where he studied harmonics and composition; already at that stage his tendency to go his own way was conspicuous. He composed a series of songs which were later performed at various concerts. After his studies in music, he studied zoology at the University of Vienna. Subsequent to obtaining his Ph.D., he accepted a post at the newly founded Institute for Experimental Biology. Here originated the succession of exciting works which centred mostly on the inheritance of acquired characters and which made his name known almost overnight in the entire scientific world. But there was also no lack of hostility and envy; the difficulties which he experienced at his habilitation as a Lecturer in Zoology made him aware that there were many who doubted his results precisely because of their originality. He subsequently attempted in a series of major experimental reports to prove the correctness of his results and the validity of his experimental methods. In a short span of time he also published a series of books for the general reader, among which his *Allgemeine Biologie* and *Das Gesetz der Serie* enjoyed surprising popularity.

His style was extremely lively, his exposition fluent, and his delivery at public lectures inspiring. There was not one among his scientific opponents who did not willingly surrender to the magic of his voice.

His remarkable linguistic gift enabled him to master the main languages of Europe to a degree which enabled him to lecture and participate in discussions in any country. Owing to his knowledge of numerous Italian dialects, he was given during the War the task of censoring the letters of Italian prisoners-of-war.

It was a painful disappointment to him that his ambition to obtain a Chair at the University of Vienna was never fulfilled. He was extremely happy when during a visit abroad he was invited to build an institute for experimental biology in Moscow under the auspices of Professor Pavlov's famous Institute, and to occupy the Chair for Genetics. He was to leave for Moscow within the next few days and to start there on October 1. The greater the shock and pain of all his friends when the news reached them last night that he had shot himself on the Schneeberg.

Neue Freie Presse—25.IX.

DR. KAMMERER—FRIENDS ON THE MOTIVES OF THE DEED

. . . His fatal decision to end his life may have been influenced by the fact that a Viennese artiste, who was close to his heart, could not make up her mind to follow him to Moscow. . . . He could not bear the thought that while his scientific ambitions were to be satisfied, his artistic and aesthetic interests would not find the same fulfilment in contemporary Moscow as they did in Vienna. He loved music and he loved women. . . .

Neue Freie Presse—27.IX.

Two days before his suicide, Dr. Kammerer visited the Soviet Legation in Vienna and with much zest gave instructions regarding the crating and transport of the scientific apparatus and machines which he had ordered for his future experimental institute in Moscow. These crates are going to be dispatched to Moscow in the near future. . . .

Der Abend, Vienna—24.IX [Socialist daily]

For many years Kammerer was close to the writer of these lines, not only as a contributor to this journal but also as a comrade in the struggle for a socialist future in mankind.

If our social and scientific establishment were to approach Kammerer's dead body, the corpse would lift its arm, as the old German legend has it, to indicate the presence of his assassins: a social order which denies an eminent scientist that secure existence which is indispensable for creative activity; a scientific orthodoxy that denied him the recognition, the means for teaching and research which are his due, only because he did not think, feel and act in an orthodox manner. . . .

*Neues Wiener Tagblatt—*26.IX.

IN MEMORIAM PAUL KAMMERER
by Peter Sturmbusch

In diesem Lande genial zu sein
Ist von der Kirche und dem Staat verboten.

(To be a genius in this country
Forbidden is by Church and State.)

The obituaries give the impression of kaleidoscopic images, as mottled and varied as the spotted salamander whose colour changes he had studied for so many years. The combination of journalese and emotion lends them a quality of contemporaneity which no biographer could hope to recapture after half a century.

4

Paul Kammerer was born in Vienna on August 17, 1880, under a lucky star, for the Emperor Franz Josef was born on August 18, 1830, so that young Paul's birthday celebrations were followed by nation-wide festivities.

The Kammerers were of Saxonian origin, but some enterprising forbears migrated to Transylvania, and their descendants to Vienna—El Dorado of Central Europe in the nineteenth century. They were a prosperous family. Paul's father, Karl Kammerer, was the founder and co-proprietor of the leading factory for optical instruments in Austria. But they were far from being conventional *bourgeois*. After twenty years of married life, Karl Kammerer divorced his wife and married Sofie, a statuesque and temperamental widow from Hungary. It was her third marriage; she brought two sons into the new household, and Karl one, all three nearly grown up. To let their newly wed parents enjoy their connubial bliss, the three boys were packed off to complete their education in England, and returned to Vienna 'more English than the English'. Charley, in particular, 'dressed like an Englishman in a cartoon'.*

Thus when Paul Kammerer was born, he had three step-brothers eighteen to twenty years older than himself. All three 'adored the new baby, and throughout the years remained loving and loyal brothers'. That was lucky: they might just as well have been bullies. Or, if they had been adoring sisters, they might have turned the little boy into a sissy. As it happened, the three stepbrothers did spoil the child, but in a sporting, masculine sort of way. Perhaps they wanted him to grow into their idea of an English gentleman. Charley was mad about dogs, collected Oriental carpets and binoculars. Lacerta remembers him 'going for a walk, all hung with binoculars like a Christmas tree'. He was an expert in training Alsatians, and gave her books and lessons in dog training. In all likelihood he had done the same, in his time, to her father, and awakened the boy's peculiar gift for establishing rapport with all kinds of animals, from dogs to snakes and lizards, down to frogs and toads. Years later, at the height of his fame, when Kammerer was a guest at a castle in Moravia, he picked up a rare variety of *Kröte*—toad— in the garden and kissed it tenderly on the head. The old chatelaine, who was present, almost fainted, and henceforth called him *der Krötenküsser*.

As the only child of ageing parents, cosseted by those three benevolent big brothers, he unavoidably grew into a *Wunder-kind*; but while prodigies when reaching adulthood often be-come emotionally thwarted, young Kammerer went all out for 'music and women', in addition to mountaineering and keeping a smelly private zoo of pets.

But he started on the wrong foot, by studying music at the Vienna Academy before reading zoology at the University. It was a bad mark against him which the Establishment never forgot. For a scientist to play the piano, as a hobby, is permissible; for a pianist to switch to science is not. It inflicts on him the stigma of dilettantism which he will never get rid of. '*Ne supra crepidam*', the robed Viennese academics would quote to each other with a knowing titter; the cobbler should stick to his last.

* Where no sources of quotations are indicated they are from Lacerta Kammerer's letters.

Music was endemic in the family. Father Karl (also a mountaineer) collected musical boxes; Mother Sofie was a piano fiend.

She was very erect, carefully dressed, and had snow-white, beautiful curls. She was slim, figure-conscious and slept in her stays. She was addicted to piano-playing. Karl was reported to have said: 'Ever since I married an automatic piano, I have lost all my love for music.' Somebody else remarked about her playing: 'What's the difference between Sofie and a sewing machine? Sofie is quicker than a sewing machine, but the latter has more feeling.' Yet she had a very good ear. She played light classics from memory, and had a large range. To show off, she squeezed in her own arabesques without quite ruining the tune. She couldn't play a thing without adding to it. I remember her when she was probably approaching eighty sitting at the piano making a tremendous noise, with her arms swinging and her fingers racing up and down.

Kammerer composed mostly songs; they were performed at concerts in Vienna, but the printed editions are untraceable. Lacerta Kammerer, however, has preserved a few manuscript scores which sound both original and charming. Their style is highly individual, reflecting the early influence of Mahler and the later influences of Schönberg and Alban Berg*. The family and their friends seem to have spent most of their evenings at concerts or at the opera. Among young Kammerer's friends before the First World War were Bruno Walter, another prodigy, who at the age of twenty-four became conductor of the Vienna Opera, and the much-admired Gustav Mahler himself, Director of the Opera and nicknamed 'the Tyrant' for the iron discipline he exacted from its members as a price for greatness. After Mahler's death in 1911, his widow, the legendary Alma, became for a while Kammerer's assistant, in charge of some experiments—rather fittingly—on the moulting habits of the praying mantis.†

* The songs were recorded after more than half a century of oblivion, for transmission on a BBC–TV programme based on this book.
† Alma Mahler-Werfel, as she was later known, delighted in the role of the *femme fatale* among the celebrities of Austria. After the Kammerer episode she had her celebrated affair with the painter Oskar Kokoshka, followed by her celebrated marriage to the architect Walter Gropius, followed by her celebrated marriage to the writer Franz Werfel, not to mention some celebrated minor affairs. In her memoirs

Vienna before the First World War is as remote as the vanished continent of Atlantis. It was a glittering world of opera, theatre and concerts, of picnics on the Danube, summer nights in the vineyards of Grinzing, and love affairs light as fluff. It was also a world of corruption and decadence, creaking in all its multi-national joints, waiting to fall to pieces. But who, at twenty, listens to the whispers of doom? Schnitzler's *Reigen* —*La Ronde*—was written when Kammerer was that age. He must have thoroughly enjoyed himself as a student at the Academy, at the University and, soon afterwards, as an international celebrity. And he certainly enjoyed, with equal zest, his work at the *Biologische Versuchsanstalt*, the Institute for Experimental Biology, known among biologists as 'the Sorcerers' Institute'. He joined it at the age of twenty-two, and stayed on until nearly the end.

5

The 'Sorcerers' were the founder of the Institute, Professor Hans Leo Przibram, and his collaborators. As he plays an important part in Kammerer's life, and in the controversy about his work, I must say a few words about the Przibrams.

They were a dynasty of scientists, originally from Prague, comparable to the Huxleys or Batesons in England, and the Polanyis from Hungary. At the beginning of our century there were no less than six Przibrams holding professorships at different faculties of the University of Vienna.[3] Hans Przibram's father, Gustav, though a politician, was one of the first citizens of Vienna to install electric lighting in the family flat, by means of a self-made battery. Hans' brother, Karl Przibram, is Professor Emeritus of Physics in Vienna, and at the age of ninety-one is a lively correspondent to whom I owe much valuable information. Hans Przibram himself was not only an eminent biologist* but of a benevolent, almost saintly disposi-

(which the editor himself describes as 'dangerously erratic' and 'impassioned') she relates that Kammerer fell madly in love with her, subsequent to a 'reluctantly granted kiss', and threatened to shoot himself on Gustav Mahler's grave unless she married him. Although she got all her facts wrong, and wrote with undisguised venom, Kammerer's infatuation and romantic threat have a ring of truth, and are in keeping with the Byronic streak in his attitude to women—about which later.
* Apart from many technical publications, he wrote the monumental *Experimental-Zoologie*, in seven volumes, published between 1907 and

tion, rare among scientists, whose life ended in martyrdom.

The Przibrams were a rich family. In 1903 the so-called Vivarium, a huge pseudo-Renaissance building in the Prater—Vienna's entertainment park—came up for sale. It had originally been a show-aquarium, to which later a 'terrarium' for reptiles and other land animals was added; thus it became the 'Vivarium'. But the Viennese were more interested in Punch and Judy shows, so the enterprise went bankrupt. Hans Przibram, together with two other well-to-do scientists,* bought the building and transformed it into the Institute for Experimental Biology.

Experimental biology was at that time a new branch of research—a revolutionary break-away from the purely theoretical and descriptive type of zoology taught at the universities. Already, as a student, Przibram had fallen under the spell of its pioneers—Wilhelm Roux and Jacques Loeb. The old Vivarium became the first scientific institute specialising in biological experimentation. It was also the first to be equipped with a primitive kind of air-conditioning, which made it possible to keep the temperature in the breeding laboratories constant anywhere between +5 degrees and +40 degrees centigrade, and to control the humidity of the air. Perhaps Przibram was inspired by the electric lighting installed by his father. No wonder the Vivarium became a centre of pilgrimage for biologists from all over the world. Among the young scientists who worked there, for shorter or longer periods, several became internationally famous later on—among them von Frisch, discoverer of the dance-language of the bees, Paul Weiss, whose limb-transplantation experiments on amphibians had a revolutionary impact on the study of the nervous system; the Hungarian Kopany, the first to transplant eyes (in rats)—not to mention Alma Mahler and her praying mantis. Some of the experimenters may have been carried away by the euphoria often found when a new branch of science opens up: one Dr. Finkler transplanted heads from male to female insects which showed signs of life for several days but, allegedly, disturbed sexual behaviour; and Professor Steinach's rejuvenation experiments by stimulating the internal secretion of the sex glands became all

1930. Cf. also Sir D'Arcy W. Thompson's obituary of Przibram in *Nature*, June 30, 1945.
* Professor Leopold von Portheim and Professor Wilhelm Figdor, both botanists.

the rage in the popular press about the time of the discovery of Tutenkhamon's tomb. These were fringe phenomena, the by-products of over-enthusiasm; but they were eagerly seized upon by the orthodoxy, who held that experimenting is a dirty job, fit for chemists, but beneath the dignity of zoologists. As a result, work published by members of the Institute was regarded with distrust and gave rise to heated controversies. Przibram's integrity, however, was never in dispute.

6

At the time the Institute was founded, young Kammerer was thoroughly fed up with the old-fashioned type of zoology taught at the University, and contemplated going back to the Academy of Music. He had already published several articles in naturalist journals, among them one about the 'Reptiles and Amphibians of the High Tatra Mountains', another about 'Amphibians in Captivity'.[4] These articles were read by Przibram just at the time when he was starting the new Institute. He got in touch with Kammerer—and thus began a life-long association:

> We were looking [Przibram wrote] for an assistant to plan the layout of the aquaria and terraria where the little creatures would feel at home. Having read Kammerer's article on the care of his animals in captivity, I paid him a visit and discovered in him an enthusiastic and competent collaborator. His gifts as a musician and his artistic temperament were matched by his competence in observing nature in minutest detail and a love of all living creatures without parallel in my experience. That was the core of his personality. He organised our aquaria and terraria in such a way that they became models for the proper care of these animals. I have hardly known anybody who could handle them as he did. This, however, turned out to be a mixed blessing; for the main point of the experimental method is that under the same experimental conditions the same results can be obtained over and again, to confirm the original experiment. But if subsequent experimenters fail in keeping the animals alive for as long a period or for as many generations as the first experimenter did, then they are obviously not in a position to test and confirm his results.'[5]

In these lines, written shortly after Kammerer's death in a review of his scientific achievements, Przibram put his finger on the principal cause of the tragedy. Kammerer was a kind of wizard with lizards, and was able to breed amphibians under artificially varied environmental conditions which nobody had succeeded in doing before or after him. His main opponent in England was the great Darwinian evolutionist, William Bateson (who coined the word 'genetics'). He kept up a running controversy with Kammerer over a period of fourteen years, denied the existence of the famous nuptial pads of *Alytes*, and yet his son (the anthropologist Gregory Bateson) remembers 'my father marvelling that Kammerer was breeding *Alytes* at all in captivity'.[6] Another of his detractors, the geneticist Richard Goldschmidt, also grudgingly admits Kammerer's special knack for breeding amphibians and reptiles:

> He was a very high-strung, decadent but brilliant man who spent his nights, after a day in the laboratory, composing symphonies. He was originally not a scientist, but what the Germans call an '*Aquarianer*', an amateur [*sic*] breeder of lower vertebrates. In this field he had an immense skill, and I believe that the data that he presented upon the direct action of the environment are largely correct.[7]

It was precisely this 'immense skill' in breeding and rearing newts, lizards and toads—the absolute opposite of amateurishness—which bedevilled the issue, because nobody was able either to confirm or to refute his experiments. Bateson denied the validity of Kammerer's breeding experiments with the midwife toad, but Bateson never succeeded in breeding midwife toads at all.* Nor did G. A. Boulenger, Curator of Reptiles at the British Museum (Natural History). His son, E. G. Boulenger, Curator of Reptiles at the London Zoo, tried to repeat Kammerer's salamander experiments, but never got the *Salamandra* to breed under the required conditions (see Appendix 3). To say it once more, the normal procedure in science, when an experiment produces controversial results, is

* Proof that he tried is a letter dated September 14, 1923, from the firm of B. T. Child—113 Pentonville Road, London, N.W.1, 'professional aquarist, collector and distributor of fish, water plants, reptiles from all parts of the world'—confirming Bateson's order for *Alytes obstetricans*. He also ordered *Salamandra* from the German firm of Enghardt in Vorwohle.[8] But Bateson never published any of his experiments with either *Salamandra* or *Alytes*.

for other researchers to repeat it under the same laboratory conditions, and thus to test, confirm or refute the original experimenter's claims. But neither Kammerer's experiments with *Salamandra*, nor with *Alytes*, have been repeated with proper care to this day.

There are both good and bad reasons for this remarkable omission. The bad reasons are the odium attached to the scandal, and the fear of ridicule or academic disrepute which such 'Lamarckian' experiments might entail. The 'good', or at least acceptable, reasons are that these creatures are extremely difficult to breed and manipulate—unless one happens to be an *Aquarianer*, and unless the experimenter is, in Kammerer's words, 'willing and prepared to devote a large part of his lifetime to the work'.[9]

This Kammerer did. The bewildering multitude of his passions and hobbies was paradoxically combined with his single-minded, dogged patience in experimental work. The hereditary changes he attempted to induce in his experimental animals could be expected to appear only after several generations; and most biologists were unable to breed even a second generation. In a lecture given in Vienna in 1914 he remarked wistfully:

Unfortunately, repeating my experiments is a difficult undertaking; they extended over a span of ten years or more; we would have to wait for at least that long for further confirmation. Moreover, the techniques of breeding— which often require much patience over a decade and over many generations, fraught with the danger of extinction of the line before the results show—these techniques have so far found too few disciples among professional zoologists; so that I am still almost the only one who practises them.[10]

In these 'ten years or more'—in fact they were extended from his early twenties to the age of thirty-five—Kammerer's experimental reports were published in rapid succession, and made biologists all over the world sit up. His most important papers were published in the *Archiv für Entwicklungsmechanik der Organismen*, familiarly known as 'Roux's *Archiv*'. It was edited by the pioneer of experimental embryology, Wilhelm Roux, and was at that period, as the *Encyclopaedia Britannica* has it, 'one of the most respected, if not the most respected, journal in the biological sciences in the world'. A number of other papers appeared in the no less respectable *Zentralblatt für Physiologie*, *Natur* (Leipzig), *Nature* (London) and other technical journals. The point needs stressing because his later detractors tried to create the impression that he was nothing but a popular journalist and amateur. Thus the late H. Graham Cannon, F.R.S., Professor of Zoology at the University of Manchester (himself a Lamarckian, but of a different brand), wrote in 1959—more than thirty years after Kammerer's death:

> The true story of Kammerer's experiments is so little known and what has been published is so partisan that I feel that it is only right that I should record what I know of the trouble. Kammerer's *Alytes* work was first published in a short paper just before the First World War.[11]

A glance at the volumes of Roux's *Archiv*, to be found in any university library, would have told Professor Cannon that the first in the series of Kammerer's papers on *Alytes* dates from 1906 and occupies ninety-two pages of solid German textbook size; the next, in 1909, occupies ninety-nine pages, the last, in 1919, forty-seven pages.[12] The description of 238 pages crammed with technical detail, published over a period of fourteen years, as 'a short paper just before the war' gives us a foretaste of the methods used by Kammerer's opponents.

Two

Before taking a closer look at Kammerer's experiments I must try to explain why they were so 'specially exciting' that they 'stirred European biology', to quote Richard Goldschmidt again.[1] The short answer is that they attempted to prove the inheritance of acquired characteristics which Lamarck had postulated. 'Acquired characteristics' in this context mean improvements in bodily structure, skills, habits, or ways of life, which the parents acquire through their efforts to cope with their environment, to adapt to its conditions and exploit the opportunities it offers. In other words, these 'acquired characteristics' are progressive changes which correspond to the vital needs of the species, and which, according to Lamarck, are transmitted to the offspring through the channels of heredity. Each generation would thus derive some benefit from the struggles and exertions of its forbears by *direct bodily inheritance* (and not merely indirectly through imitative learning from its elders). To put it crudely, the blacksmith's son would be born with stronger than average biceps, without having to develop them by repeating his father's efforts all over again; and Miss Europe's daughters may be born with a slimmer waistline without having to starve all over again. This is overstating the case, because Lamarckians generally believe that only essential characteristics which have been acquired in response to intense and persistent challenges of the environment over several generations become eventually inherited. But, nevertheless, the essence of Lamarckism is the belief that the efforts of the parents are not entirely wasted, that some of the

27

benefits derived from them are transmitted to their offspring; and that this is the principal active cause of evolution from amoeba to man. Therein lies the great philosophical attraction of this view, which can be traced back as far as Hippocrates. In a popular lecture on 'The Significance of the Inheritance of Acquired Characteristics for Education' Kammerer made eloquent use of this appeal.

> Fröbel, Pestalozzi [educational reformers much in vogue at the time] and their schools relied on the potential dispositions which the child inherits from its ancestors, hereditary dispositions which the educator hoped to enrich; but he could not hope to bestow on the children a permanent heirloom in which their children's children would be able to share—only a gift for the fleeting duration of an individual existence. They could not conclude otherwise but that at the death of an individual his acquired merits would also die with him; his heirs might continue what the ancestor began, but however excellent their hereditary dispositions, they again had to begin at the beginning.
>
> However, on the hypothesis of the inheritability of acquired characters, which seems to be closer to the truth, the individual's efforts are not wasted; they are not limited by his own lifespan, but enter into the life-sap of generations. It depends on us whether it will produce a benign or destructive effect.
>
> By teaching our children and pupils how to prevail in the struggles of life and attain to ever higher perfection, we give them more than short benefits for their own lifetime—because an extract of it will penetrate that substance which is the truly immortal part of man.
>
> Out of the treasure of potentialities contained in the hereditary substance transmitted to us from the past, we form and transform, according to our choice and fancy, a new and better one for the future.[2]

It should be borne in mind that this lecture was addressed to an audience of schoolteachers and educationalists; Kammerer's scientific papers are written in the orthodox, dry-as-dust style. Even so, these popular lectures—and their immense success—were looked upon with a jaundiced eye by the Establishment. But just because of their popular, colourful language they help us to understand why the nuptial pads of the midwife toad—

acquired by the 'energy and diligence' of the parent toad and transmitted for the benefit of future generations—caused such an almost hysterical excitement among geneticists. This will become even clearer if we contrast the Lamarckian thesis of the inheritance of acquired characteristics, with its antithesis, the neo-Darwinian theory.

According to the Lamarckians, evolution progresses stepwise, in the commonsense manner in which a bricklayer builds: each generation profits from the accumulated experience of its forbears. The neo-Darwinian theory, on the other hand, postulates that the parents can transmit through the channels of heredity only what they have inherited themselves and nothing else— none of the new acquisitions in skills or bodily features that they have made in their lifetime. One might compare this doctrine to a law which decrees that a man can leave to his heirs only what he himself had been left by his parents, neither more nor less—not the added wealth he had acquired, not the house he had built, not the patents of the inventions he made; nor the debts he had incurred. In so far as his offspring is concerned, he might say with *Ecclesiastes* that all his efforts amounted to 'vanity and chasing the wind'. The genetic endowment is transmitted down the generations unaffected by anything that happened to its transient carriers in their lifetime. This doctrine of the 'continuity and unalterability of the germ-tract', postulated by the German zoologist, August Weismann, in 1885, was a cornerstone of Darwinism in Kammerer's time, and it still is at the time of writing. The textbooks tell us that the genetic blueprint is located in the chromosomes of the germ-cells, which are kept in splendid isolation from the rest of the body. They are potentially immortal molecular structures, protected from the hazards of life, and passed on, unaltered, from generation to generation along the 'continuous germ-tract'. In the Lamarckian view, evolution is *cumulative*; in the Darwinian, *repetitive*; it could go on for millions of generations without any evolutionary progress.

How, then, did the blueprint for amoeba nevertheless become transformed into the blueprint for man? According to neo-Darwinian theory, this happened thanks to the occasional ocurrence of chance events on a microscopic scale, called 'random mutations'. Mutations are defined as spontaneous changes in the molecular structure of the chromosomes; and they are said to be 'random' because they are supposed to be totally unrelated

29

to whatever goes on in the animal's environment, and thus totally unrelated to its evolutionary needs. Since they are random, most of these mutations produce harmful or lethal effects, but there must also occur from time to time a few lucky hits which confer some minute advantage on the carrier of the mutated chromosome, and this will be preserved by the operation of natural selection.

The above is a necessarily simplified summary of the two opposite theories. But it should help to explain the intense emotional and philosophical passions which the controversy evoked. As Sir Alister Hardy said in his Gifford Lectures: 'There can be no doubt that the whole concept of Lamarckism is fraught with emotion. A degree of prejudice occurs, of almost equal intensity I believe, *on each side* of the Lamarckian argument. On one side are those who are so shocked by the materialism of the Darwinian conception that . . . they cannot bring themselves to believe that it is the real mechanism of our creation. For them Lamarckism with the inheritance of acquired characters may seem the only alternative. They have this deep-felt wish to believe Lamarck to be right. On the other side are those who feel that Lamarckism is nothing but a superstition.'[3]

Neo-Darwinism does indeed carry the nineteenth-century brand of materialism to its extreme limits—to the proverbial monkey at the typewriter, hitting by pure chance on the proper keys to produce a Shakespeare sonnet—since, as Sir Julian Huxley once said, 'given sufficient time, anything at all will turn up'. It is by no means only Lamarckians who find this conception difficult to accept; there is a considerable proportion, perhaps even a majority, of eminent biologists inside the scientific Establishment, who reject Lamarckism and yet feel that while the Darwinian theory of natural selection operating on random mutations answers *some* of the problems posed by evolution, it leaves the most important ones unanswered.*

Added to these logical difficulties, there is a metaphysical flavour attached to Darwinism which not only the followers of Bishop Wilberforce found repugnant: the concept of 'blind

* Thus, for instance, Professor Waddington, who would sue for libel if one called him a Lamarckian, has compared the theory of evolution by chance mutations to 'throwing bricks together in heaps' in the hope that they would 'arrange themselves into an inhabitable house'; and as for natural selection, it 'in fact merely amounts to the statement that the individuals which leave most offspring are those which leave most offspring. It is a tautology.'[4]

chance' as the universal law of Nature, on which Einstein made his famous comment, 'I refuse to believe that the Creator plays dice with the world'. At the same time, the doctrine of the continuity of the germ-track, of a fixed genetic endowment of the unborn child, which nothing can alter, suggests the idea of a mechanistic sort of predestination. When William Bateson was in the First World War lecturing to the troops on Darwinism, a soldier commented: 'This is scientific Calvinism.' Bateson called this 'a flash of illiterate inspiration'.

Perhaps the profoundest reason for opposing Darwinism is contained in a remark by Henri Bergson, whose vitalistic philosophy was directed against the mechanistic trend: 'The vitalist principle may indeed not explain much, but it is at least a sort of label affixed to our ignorance, so as to remind us of this occasionally, while mechanism invites us to ignore that ignorance.'[5] But Bergson's restraint was exceptional. More revealing is this outburst by the Manchester Professor of Zoology, writing in 1959, after the Darwin centenary celebrations:

> At these meetings, with their adjuncts of broadcasts and publications, all my criticisms were completely ignored. But what was much worse, the name of Lamarck was introduced in a manner that is scarcely possible to believe would occur in an enlightened scientific world. Despite the fact that some years ago I published a paper 'What Lamarck Really Said', in which I pointed out some of the grosser ways in which the views of Lamarck had been misrepresented in the middle of the last century and the beginning of this . . . the pundits who pulled the strings behind the celebrations persisted in repeating in a most aggressive manner the calumnies that have been perpetrated against Lamarck.
>
> Now things cannot go on like this. The orthodox geneticists are certainly well entrenched. But they are bound ultimately to face up to the mounting barrage of opposition which is being published against them. They cannot forever protect themselves in their trenches of cabbalistic formulae.[6]

Two generations earlier, about the time when Kammerer was born, a famous Lamarckian, Samuel Butler, wrote in his *Notebooks*: 'I attacked the foundations of morality in *Erewhon*, and nobody cared two straws. I tore open the wounds of my

Redeemer as he hung upon the Cross in *The Fair Haven*, and people rather liked it. But when I attacked Mr. Darwin they were up in arms in a moment.'[7] In his *Evolution Old and New* (published in 1879) Butler added: 'Lamarck has been so systematically laughed at that it amounts to little less than philosophical suicide for anyone to stand up on his behalf.'[8]

This was written nearly fifty years before Kammerer committed suicide. Another thirty years later a prominent member of the Establishment (Sir Gavin de Beer) called 'attempts to impugn' Darwin's teaching an 'exhibition of ignorance and effrontery'. And another (Professor Darlington) called the Lamarckian theory 'a disreputable and ancient superstition'.[9] A controversy which goes on for nearly a century with such unabated fury must have very deep emotive roots.

Ironically, Darwin himself did not share the view that Lamarck was disreputable. On the contrary, in his early notebooks, not intended for publication, he paid tribute to Lamarck as a source of inspiration 'endowed with the prophetic spirit in science, the highest endowment of lofty genius'. Later on he changed his mind, and in his letters referred to Lamarck's theories as 'veritable rubbish'.[10] But then he once more changed his mind, and in his *Variation of Animals and Plants under Domestication*, published in 1868, he gave a whole series of alleged examples of Lamarckian inheritance of acquired characteristics—horses inheriting bony growths on their legs which their parents had developed through travelling on hard roads, a man losing part of his little finger and all of his sons being born with deformed little fingers, and similar old wives' tales which he believed to be gospel truths.* Four years later he inserted a newly written chapter in the sixth edition of *The Origin of Species* on the same lines; and another three years later, in a letter to Galton,[12] he admitted that each year he found himself more compelled to revert to the inheritance of acquired characteristics—because chance variations and natural selection alone were apparently insufficient to explain the phenomena of evolution. The examples that he quoted were no doubt apocryphal,

* Darwin also wrote: 'Circumcision is practised by Mohammedans, but at a much later age than by Jews; and Riedel, assistant resident in North Celebes, writes to me that the boys there go naked until from six to ten years old; and he has observed that many of them, though not all, have their prepuces much reduced in length, and this he attributes to the inherited effects of the operation.'[11]

but they prove that if Lamarckism was a superstition, Darwin shared it.

Why, then, all the sound and fury? Apart from the reasons already given, disciples tend to be more fanatical than their masters; they have committed themselves to his system, invested years of labour and staked their reputation on it; they fought the opposition and cannot tolerate the idea that the system might be at fault. To out-Herod Herod is a phenomenon as common among scientists devoted to their theory as it is among politicians or theologians devoted to a doctrine—whether Freudian or Jungian, Stalinist or Trotskyist, Jesuit or Jansenite.

Moreover, to all appearances, Darwinism offered a 'modern', mechanistic explanation of evolution, which Lamarckism was unable to do. The discovery of Mendel's Laws, the statistical approach to genetics, and finally the breaking of the 'genetic code' imprinted on the chromosomes, seemed to be as many confirmations of Darwin's prophetic foresight. The mechanism of evolution which he had proposed may have been crude, in need of modifications and refinements; but the Lamarckians could offer no mechanism at all which would be in keeping with modern biochemistry. Random mutations in the chromosomes, triggered by radioactivity, cosmic rays, excessive heat or noxious chemicals, were scientifically acceptable as a basis on which natural selection could operate. But no acceptable hypothesis was forthcoming to explain how an acquired bodily or mental feature could cause an alteration in the genetic blueprint, contained in the micro-structure of the chromosomes in the germ-cells. That evolution should operate through a process which permits the offspring to benefit from useful changes in its forbears was an idea that might appeal to common sense, but to the scientist at his microscope it was technically unimaginable and had to be rejected. It smacked of the ancient notion of a miniature homunculus, encased in the sperm or ovum as an exact replica of the person who carried it—including his or her 'acquired characteristics'; so that the individual who grew out of the homunculus would bear the mark of everything that had happened to its parents.

Thus Lamarckism acquired the stigma of a 'disreputable, ancient superstition', because it postulated a principle in nature without being able to offer a mechanism, in terms of contemporary science, which would account for it. This situation, however, is not new in the history of science. When, half a century before Newton, the German astronomer, Johannes Kepler, suggested that the tides were caused by the moon's attraction,* Galileo contemptuously dismissed the idea as an 'occult fancy' because there was no conceivable mechanism which could explain action-at-a-distance (nor is there today).

2

Universal gravity was rejected as magic because, in Newton's own words, it meant 'grappling with ghost fingers at distant objects',[13] and thus contradicted the laws of mechanics; and Lamarckism was rejected because the suggestion that the organism could interfere with the structure of its chromosomes contradicted the 'laws of genetics'. Nevertheless, the attraction of the idea was still so strong in Kammerer's day that efforts to prove Lamarckian inheritance by experiment continued, in spite of the handicap that there existed no theory to explain how it worked—if it worked. And, scandalous to report, the staunchest Darwinians were the first to indulge in a guilty spree of Lamarckian experimentation. Darwin himself did not have to do so—he was, as we saw, easily convinced by hearsay. While working on *The Origins*, he wrote in his notebook: 'The cat had its tail cut off at Shrewsbury. Its kittens had all short tails; but one a little longer than the rest; they all died. She had kittens before and afterwards with tails.' And in his later works he gave even more scurrilous examples.

August Weismann, more Darwinian than Darwin, had postulated that the substance which carried the hereditary dispositions—the 'germ-plasm', as he called it—remained unaffected by acquired characters. One of his famous experiments was to cut off the tails of mice, for twenty-two successive generations, to see whether a tailless mouse would appear. But, as a Lamarckian critic remarked, he might as well have studied the inheritance of a wooden leg. For Lamarck's thesis was that only such characteristics are inherited which an animal develops as a result of its natural, adaptive needs—and losing its tail by amputation could hardly be called an adaptive need of the mouse.

In the Soviet Union where, as we remember, the Party line was pro-Lamarck, the great Pavlov himself, originator of the

* In the Preface of Kepler's *Astronomia Nova*, published in 1609.

conditioned-reflex theory, attempted to show in his Leningrad laboratory that the results of conditioning were inheritable. He trained mice to respond to a bell which was sounded each time before the arrival of food. In the first generation, the mice needed three hundred 'lessons' to learn that the bell signalled food; the second generation needed only one hundred lessons; the third, thirty; and the fourth cottoned on to 'bell signals food' in a mere five lessons. Here, it seemed, was proof of the inheritance of knowledge acquired by learning—and a method to produce supermen by conditioning over several generations. But when the experiment was repeated, and brought negative results, Pavlov publicly withdrew his claim, explaining that it had been based on faulty experiments by an assistant (the laboratory assistant as a culprit will turn up repeatedly in these pages). Pavlov's disclaimer proved his integrity as a scientist; yet he retained his faith in the inheritance of acquired learning, for the year after the abortive mouse-experiments, Kammerer was invited to build a biological laboratory in Russia, affiliated to Pavlov's Institute.

A famous psychologist, William MacDougall, Professor at Harvard, also tried his best to demonstrate Lamarckian inheritance. He trained rats to choose the proper escape route from a tank full of water, bred the trained animals and subjected their progeny to the same training. As with Pavlov's mice, the rats of each succeeding generation needed fewer and fewer lessons to learn the escape route. When he published his results, in 1927, he did not strike a note of triumph, but rather of diffidence, speaking with remarkable frankness:

> In this connection it is necessary to avow that, during the course of the experiment, there grew up in all of us a keen interest in, I think I must in fairness say, a strong desire for, positive results. . . . On my own part, there was a feeling that a clear-cut positive result would go far to render tenable a theory of organic evolution, while a negative result would leave us in the Cimmerian darkness in which Neo-Darwinism finds itself.
>
> I was conscious, therefore, of a strong bias in favour of a positive result; and throughout I was consciously struggling against the temptation to condone or pass over any detail of procedure that might unduly favour a positive result. Such details are encountered at every point, more especi-

ally in the breeding of the animals. To have disguised from oneself this bias, to have pretended that we were superior to such human weakness, would have been dangerous in the extreme; the only safeguards against its influence were the frank avowal of it and unremitting watchfulness against it. . . . I can only say that I believe we have succeeded in standing upright; and in fact, for myself, I am disposed to believe that I have leaned over backwards, as we say in America. Whether we have really succeeded in this, the most difficult part of our task, can only be proved when other workers shall have undertaken similar experiments. If our results are not valid, the flaw, which has escaped our penetration hitherto, must, I think, be due to some subtle influence of this bias.[14]

His dark premonitions proved to have been correct. The experiments were repeated (by Professor Agar and his team in Melbourne) with dogged patience over fifteen years. They confirmed MacDougall's claim that the progeny of trained animals learnt quicker than their parents had done—but, alas, the control animals of untrained ancestry *also* learned quicker than their parents. Apparently, rat stocks bred in laboratory conditions improved in intelligence over the generations.* Once more the verdict could only be: not proven.

'The story of the Lamarckians,' Sir Alister Hardy comments, 'has, I think, been a particularly sad one. By intuition they feel certain that changes in behaviour have played a much greater part in evolution than their colleagues will admit. In this I am sure they are right, but they have failed to provide a convincing argument to support their case.'[15]

To round off these remarks on the great controversy which provided the background to Paul Kammerer's tragedy, I shall quote a passage by that perhaps most passionate anti-Lamarckian, Professor Darlington, in his *Facts of Life* (compressed):

Down the ages we have seen a cleavage always separating two theories of heredity. The first was the old theory, the direct or vitalistic theory, that the parent in his whole character at the time of begetting was represented in the

---

* This is quite in keeping with the results of recent experiments undertaken, for a different purpose, by David Krech and his team in Berkeley.

offspring. Each generation was supposed to be somehow condensed into a seed or an egg from which it expanded into the next generation. The offspring therefore bore the marks inflicted by nature on the person of the parent. Acquired characters were inherited.

The second theory, the new theory, the *indirect* theory, was more difficult. It was represented in the notion of the *genitalia corpora* of Lucretius. It was expressed in the repudiation of the inheritance of acquired characters. It assumed that something which was passed down from generation to generation determined the character of the body but was itself uninfluenced in its character by what happened to the body. This theory was espoused for the first time in an unequivocal form by Weismann. It was obviously supported by the chromosome theory as it took shape in the hands of the German and American cytologists. And it fitted the new teachings of Mendelism like a glove. Indeed every improvement in the design and interpretation of experiments made the direct theory more difficult and the indirect theory more plausible.

At the beginning of our century the ancient cleavage between the two theories of heredity thus became, for the first time, a cardinal issue in the study of life. On the one hand, where the impact of genetics and cell study were felt, a change was taking place. The old idea that personal adaptations, the peculiarities forced upon us by lucky or unlucky circumstances, or by an ameliorative purpose in the Creator, or by the power of the will, were inherited was giving way. The new notion of hard particles, microscopically visible and mathematically predictable, incorrigibly deterministic and resistant to the interference of any divine purpose apart from that reflected in natural selection, was taking its place.

Where the impact of genetic experiments was not felt, however, older views still held their ground. The educated man, the classical scholar and the economist, the psychologist and the jurist, the historian, the social scientist, and the liberal philosopher, universally, and sometimes enthusiastically, believed in the inheritance of acquired characters. It made progress so much easier. There was therefore a great public demand for a reply to the uncompromising and apparently unrefined doctrine of genetics.

This unusual tension in the world of ideas led to a strange response. . . .

It happened that in 1904 a plausible young zoologist found his way into the Biological Research Institute in Vienna. His name was Paul Kammerer. He showed himself to be expert in breeding and rearing all kinds of frogs and toads, lizards and salamanders. . . .[16]

Three

I

The following is a brief and simplified account of Kammerer's
most important experimental work. All of it was based on intricate experimental designs, using a variety of different controls,
and often combining different techniques; but I shall have to
confine myself to the essential purpose of each experiment, and
must refer readers with a special interest in the subject to the
appendices and bibliography.

Kammerer collected most of his specimens on long, lonely
walking tours in the mountains and lowlands of Central
Europe, and also on three expeditions to uninhabited Dalmatian
islands. Another expedition, with Przibram, took him to the
Sudan, to collect tropical specimens for the Institute. He did
not like buying animals from the dealers because they were
'spoilt': starved or overfed, neurotic, often unwilling to mate.
He regarded reptiles and amphibians as sensitive, delicate
creatures; in his last book there are long, charming descriptions
of the varieties of temperament among lizards on the Dalmatian
isles.

The general design of his most famous experiments was to
induce modifications in the mating habits, colour or physique
of his animals by breeding them in an environment radically
different from their natural habitat (e.g. different climate, or
black instead of yellow soil), in order to find out whether these
adaptive modifications would become hereditary. His contention
was that they did—sometimes in a single generation, but in
other cases only after five or six generations, bred in the artificial
environment.

39

2

In his first sustained series of experiments he showed that the method of propagation of *Salamandra* could be altered. This work took five years to accomplish, and brought him in 1909, at the age of twenty-nine, the much-coveted Sömmering Prize for Fundamental Discoveries in Physiology, awarded by the Natural Science Society of Frankfurt, after a member of the Society, Professor Knoblauch, had confirmed Kammerer's results.

There exist in Europe two species of these long-tailed, newt-like amphibians: the black *Salamandra atra*, which inhabits the Alps, and the spotted *Salamandra maculosa*, which inhabits the lowlands. Once or twice a year the female of the spotted salamander gives birth to ten to fifty small larvae, which she deposits in the water. The larvae are tadpoles with external gills; only after several months do they shed them and their other tadpole attributes, and metamorphose into salamanders. The Alpine species, on the other hand, gives birth on the dry land to only two, fairly large and fully formed salamanders; the larval stage is absolved in the uterus.

Essentially, Kammerer's experiment consisted in breeding the spotted salamander in a cold and dry, imitated Alpine climate, and *vice versa*, breeding the Alpine salamander in an artificially hot and moist lowland climate. The results, published in communications in the *Archiv für Entwicklungsmechanik* and *Centralblatt für Physiologie*, 1904[1] and 1907,[2] showed, in his own words, 'a complete and hereditary interchange of reproductive characters'. The spotted, lowland salamander transposed into a quasi-highland climate with no rivers to breed in, eventually (after several abortive litters of tadpoles) gave birth to two fully developed salamanders, as the Alpine salamander is wont to do. And the Alpine salamander, transposed into a hot and humid climate, deposited its young in the water, instead of on land; the young were tadpoles instead of adult forms; and there were successively more and more of them with each litter. This in itself was a remarkable *tour de force*, as even Richard Goldschmidt admitted.

The second, critical, step in the experiment was to rear these 'abnormally' born specimens to maturity, let them mate, and see whether the second generation showed signs of having inherited the abnormal reproductive behaviour of the parents. This took several years, for salamanders do not become sexually mature before the age of four years (though in captivity somewhat earlier). Starting at the beginning of 1903 with a stock of forty 'abnormally' born salamanders of both sexes, he had the satisfaction of witnessing in the course of 1906 and 1907 the birth of six litters—four of the spotted, two of the Alpine type. All of them showed the artificially induced reversal of reproductive method in varying degrees.* Kammerer concluded that the experimental data demonstrated 'the inheritance of acquired characteristics as plainly as one might desire'.[3]

When Kammerer, at the age of twenty-two, started on his salamander experiments he was not motivated by the wish to prove Lamarckian inheritance. 'I was then', he wrote, 'under the spell of Weismannism and Mendelism, which both agree that acquired characters are not inherited.'[4] It was only when he saw what drastic and purposive adaptations of appearance and behaviour he was able to produce by altering an animal's environment that the idea occurred to him to test whether these useful changes might be inheritable. Commenting on the results that I have just described, he wrote: 'The whole experiment was carried out long ago by Nature herself, and has given rise to a type of salamander which inhabits high mountainous areas where the water-courses necessary for depositing tadpoles are lacking: the black Alpine salamander, *Salamandra atra*, is apparently a derivative form of our spotted salamander, which goes through its tadpole stage in the mother's womb . . . and thus perseveres until it is born.'[5]

3

The spotted salamander is a versatile little beast. According to circumstances, as we saw, it gives birth in the water or on land, to larvae or to young adult forms. Moreover, the spotted type will also change its colour, like the chameleon; but—unfortunately for the experimenter—at a much slower rate. The variety

---

* This variation in degrees was Bateson's main criticism of the results. But salamanders are not the only animals whose reproductive patterns show a considerable range of variations, and if Kammerer had intended to doctor his results, he would have smoothed over the differences rather than describing each litter in page-long, inordinate detail.

which Kammerer collected in the Vienna woods (*Salamandra maculosa, forma typica*) has yellow spots irregularly distributed on a black ground.* The preliminary stage in the next series of experiments to be described—which ranged over eleven years —was to rear one group of animals on black soil, another group on yellow soil. In the first group, the yellow spots gradually diminished until, at full maturity around the sixth year, they were quite small. In the second group the yellow spots expanded and merged into large stripes. This part of the experiment was repeated by others, and is not contested. It is also established that these colour changes were not caused by the direct chemical or photo-chemical action of the environment on the animal's skin; they were mediated by its central nervous system, reacting to the colour perceived through the animal's eyes.†

As in the case of the previously described experiment, it was the second step which counted: the attempt to prove that these adaptive colour changes of the first generation had an influence on heredity. If Kammerer's technical papers, photographs and drawings are accepted as evidence, they definitely had. The progeny of black-adapted parents, again reared on black earth, were born with a single row of small yellow spots along the median line on the back, which became smaller, and later all but disappeared. The progeny of yellow-adapted parents, again brought up on yellow soil, were born with two symmetrical rows of yellow dots, which later united into two broad yellow stripes; and the third generation brought salamanders whose back became of 'uniform canary-yellow colour'.[6] Years later, in a book intended for the general public, Kammerer wrote: 'Completely unexpected to myself was the rearrangement of the design in the second generation from irregular to symmetrical spots or stripes. The two experimental series [black-adaptation and yellow-adaptation] complement each other so beautifully that they fill one with admiration for the precision with which living substance reacts. If one were to paint on the progeny of one series all yellow surfaces black and *vice versa*, one would

* See plate.
† Earlier experiments had established that plaice and other flat fish, which change their colouration according to the colour of the ground on which they rest, failed to do so when blinded. For *Salamandra* Kammerer proved his point by skin-transplant experiments, and Przibram and others confirmed his results on blinded animals. Kammerer apparently could not bring himself to use that simple and cruel method.

approximately get the appearance of the progeny from the other series.'[7]

I shall not go into the further elaboration of the experiment, which led to hair-raising complications: for instance, he brought up half of the offspring of yellow-adapted parents on black soil, and of black-adapted parents on yellow soil, so that if there was colour inheritance from the parents it came into conflict with the adaptive tendency of the offspring. He also made intricate cross-breeding and ovary-transplant experiments, attempting to show that the hereditary changes followed Mendel's Laws, and to achieve 'a reconciliation of Lamarckism with Mendelism'.

Whatever the interpretation of Kammerer's results, these were pioneering experiments which rightly 'stirred Europe's biologists'. Thus one would have expected that scores of research teams would have been eagerly engaged in following up the new trail. Nothing of the sort happened; no serious attempt was made to confirm or refute them.

4

The greatest sensation among all of Kammerer's experiments was created by his work with that small ungainly creature, the midwife toad, *Alytes obstetricans*.

While most other toads, and frogs, mate in the water, *Alytes* mates on land. During the mating procedure in water, the male toad clasps the female round the waist and keeps her in his grasp for a considerable time—sometimes for weeks—until she spawns her eggs, which he then fertilises with his sperm. To get a firm grip on the female's slippery body in the water, the male toad develops in the mating season swellings on its palm and fingers of a blackish colour, from which small horny spines protrude: these are the famous nuptial pads. But the midwife toad, mating on land, where the female's skin is comparatively dry and rough, does not need and does not possess these pads. The female emits a multitude of eggs attached to long strings of jelly, and the male, after having fertilised them, winds these beads of eggs round his hind-legs and carries them with him until the young hatch—hence the name midwife toad.*

Kemmerer's claim was that by inducing *Alytes* to copulate in water, like other toads, for several generations, they eventually developed nuptial pads as an acquired hereditary feature.

* See plates.

43

It could, of course, be objected—as he himself was the first to point out—that the species *Alytes* had descended, in the distant past, from ordinary water-mating toads equipped with pads. Thus the reappearance of the pads would be an atavism—not so much the acquisition of a new feature as the reappearance of an old one. In a memorable lecture to the Cambridge Natural History Society in 1923, Kammerer commented: 'As the atavism objection can always be raised, it is not very clear to me why just this experiment [with *Alytes*] is so often looked upon as an *experimentum crucis. In my opinion it is by no means a conclusive proof of the inheritance of acquired characters*' (my italics).[8] Nevertheless, the hereditary fixation of an acquired, adaptive—or re-adaptive—feature was a puzzling phenomenon, and seemed difficult to reconcile with the 'no interference with the germ-track' theory. Thus, paradoxically, while Kammerer himself belittled the importance of the nuptial pads and denied that they were conclusive proof for the inheritance of acquired characteristics, his opponent Bateson *did* regard this as the *experimentum crucis*, the 'most astonishing'[9] among Kammerer's experiments, the 'critical observation . . . if it can be substantiated, it would go far to proving Kammerer's case'.[10] As one commentator wrote, the salamander experiments were so much more remarkable, that to pass them in silence and to get excited about the nuptial pads was like 'straining at a gnat and swallowing a camel'.[11]

As in the earlier experiments with the Alpine and the spotted salamanders, the first step consisted in altering the midwife toad's method of propagation. By keeping the toads in an abnormally high temperature (25–30 degrees centigrade) and providing them with a basin of cool water, the animals were made to spend more and more time in the basin, and eventually took to copulating in the water. But the eggs which the female ejected in the water swelled up at once and did not stick to the male's legs; they sank to the bottom of the basin, where most of them perished. It was again a *tour de force* on Kammerer's part to save a few of these 'water-eggs' and breed water-begotten *Alytes* from them. As already mentioned, Dr. Boulenger, Curator of Reptiles at the London Natural History Museum, and Bateson's main ally, attempted to repeat the experiments, but was unable to do so, and never succeeded in breeding a single water-begotten specimen. The reasons why he failed are discussed in Appendix 3.

44

Kammerer's discovery, in 1909, of the nuptial pads in the descendants of his water-mating *Alytes* was accidental, and he did not attach much importance to it. Later (1919), in his main paper on the subject, he wrote:

The most curious feature in the water-breed of *Alytes* are the forelimbs of the sexually mature male. It was in the F_3 generation [great grandchildren of the original couple which was induced to copulate in water] that I first noticed on the upper, outer and palmar sides of the first finger a swelling of greyish-black colour. The whole frontal extremity appeared to be more muscular and differently poised, i.e. more inward bent. When I noticed these conspicuous aberrations in generation F_3, I also found a fainter indication of them in generation F_2 [the grandchildren of the first 'water breeders']: they also showed calosities in the skin of the thumb region, but without a change of colour visible to the naked eye.

In my 1909 publication I only gave a macroscopic description and drawings of this horny calosity which is not found in normal *Alytes* males. At that time it seemed to me that this description was for the time being sufficient to satisfy any interest in the matter. Since then, however, circumstances have arisen which make it seem desirable to make a closer analysis of the nuptial pads—which is the subject of the present paper.[12]

The 'circumstances' he mentions obviously refer to the attacks by Bateson and others, which in the final part of this long paper Kammerer attempted to refute.

5

To the end of his life Kammerer believed that the 'crucial experiment' which established the inheritance of acquired characteristics was neither the work with *Salamandra*, nor with *Lacerta* or *Alytes*, but with the primitive Ascidian sea-squirt, *Ciona intestinalis*, which lives on the sea bottom and has two tube-like extensions, or siphons, waving in the sea above it, one for the intake of sea-water, the other for expelling it. Kammerer cut the syphons and found that *Ciona* replaced them by longer tubes; and the more often he repeated the operation the longer the regenerated siphons became, until they resembled 'monstrous long elephant trunks'. He then claimed that these elongations of the tubes became hereditary.

Now the elongation of the regenerated tubes (though not its hereditary character) had in fact been shown by the Italian zoologist Mingazzini, in Naples, as far back as 1891; it had been confirmed by Jacques Loeb, and nobody doubted his results. Kammerer actually got the idea of experimenting with *Ciona* from Mingazzini. But when Kammerer published his own paper on *Ciona*, the elongation of the regenerated tubes itself was immediately questioned (regardless of the question of heredity) and Mr. Munro Fox, working at the Biological Institute in Roscoff, Brittany, wrote in a letter to *Nature* that he had tried but failed to produce elongation of the regenerated tubes. Kammerer pointed out that Fox had used a wrong technique, but to no avail; whenever the subject was raised again, Munro Fox was quoted in rebuttal, but Mingazzini and Loeb were passed over in silence (see Appendix 5). Nor did anybody bother to investigate Kammerer's specimens of *Ciona*, which he had displayed in 1923 at lectures in Cambridge and London. —The original photographs of the specimens are reproduced facing p. 81.*

The point to be emphasised is that the controversy was not concerned with Kammerer's claim that the elongated siphons were *inheritable*, but got bogged down on the question whether elongated siphons occurred *at all*. At the time of writing, nearly forty years later, the matter still rests there. Yet to repeat Kammerer's experiment with *Ciona* would need neither the patience nor the delicate techniques required for breeding amphibians.

I have given a summary description of Kammerer's most publicised experiments. There were others. He worked with lizards—his beloved *Lacertae*—and claimed to have produced inherited colour changes and breeding-habit changes similar to those obtained in *Salamandra*. He experimented with the blind cave-dwelling newt *Proteus*, whose minute, rudimentary eye-buds are buried under the skin, and degenerate as the animal matures. Exposure to *normal* daylight does not restore the animal's eyesight because the skin over the eye rudiments acquires a black pigment which arrests the development of the eye and leads to the same degenerative process as occurs in

* The photographs have been preserved by Professor and Mrs. Thorpe who kindly lent them to me.

46

normal animals growing up in darkness. But by exposing *Proteus* to *red* light—which does not cause pigmentation of the skin—Kammerer produced specimens with large, perfectly functional eyes. One of his admirers, the Professor of Zoology at Imperial College, wrote later on: 'Whatever hesitation may remain about accepting Kammerer's results in other matters, there can be no dubiety about his results on *Proteus*. In common with other zoologists who attended the special meeting of the Linnean Society in May 1923, I saw these large-eyed specimens of *Proteus*, the most wonderful specimens in my judgment which have ever been exhibited to a zoological meeting.'[13]*

Kammerer did not claim that this experiment had anything to do with the inheritance of acquired characteristics; and his discovery that red light restored the vision of *Proteus* is now in every textbook. Nevertheless, although there can be 'no dubiety' about this fact, there still remained scope for perfidious insinuations. In his book on *The Material Basis of Evolution*, published in 1940, Professor Richard Goldschmidt wrote: 'Experimenting on the blind newt *Proteus*, Kammerer (1912) obtained large open eyes in individuals raised in yellow [*sic*] light.' To this sentence the following footnote is appended: 'As many of Kammerer's claims are under suspicion, I may say that I have seen the specimens. Of course, I could not swear that the good eyes had not been transplanted into the specimens.'[14]

Nine years later Goldschmidt wrote in *Science*: 'I do not believe that Kammerer was an intentional forger.'[15] But that did not erase the intentional smear. 'The eye of *Proteus*', as we shall shortly see, caused yet another scandal, with ghastly consequences for Kammerer.

* See plate.

Four

I

Kammerer joined Przibram's Vivarium in 1903 when he was twenty-three. He got his doctorate in 1904 and was appointed *Privatdozent* or Lecturer at the University two years later. In 1905 he met at an amateur concert the young Baroness Felicitas Maria Theodora von Wiedersperg, who sang several *Lieder*. 'Apparently it was love at first sight and they soon became engaged, and after the customary one year they got married in 1906. They nearly always gave the impression of being a very happy couple and they were usually most affectionate with each other.'

The Wiederspergs were a very ancient family, who traced their ancestry back to a knight crusader in the thirteenth century. But they were not snobs. Felicitas' father, Gustav von Wiedersperg, was a Member of Parliament and practised as a country doctor on his estate in Bohemia (with no charges, and the medicine given free to the peasants); one of her brothers was also a doctor, the other a civil servant. The Wiedersperg family shared the Kammerer family's passion for music and animals. Papa Wiedersperg, Lacerta writes, 'kept snakes, two monkeys, a bat, a squirrel and a stork. The Rhesus monkeys became unmanageable after his death and were taken to the Schönbrunn Zoo. I remember the stork called Mausa. When I put my arm round her, she rested her beak on my head. When she was hungry, she clapped her beak under the kitchen window.'

Thus Kammerer had married not only a pretty wife but a

48

whole clan of kindred souls, complete with their music and their pets, and soon the two families shared the Wiederspergs' house in the elegant suburb of Hietzing. Lacerta was born in 1907 and remained the only child. She, too, was soon allowed to keep her private zoo. The house, as one frequent guest* described it, 'simply stank of animals'. Felicitas was small, pretty and extremely lively, given to making outrageous pronouncements with an innocent baby face. She might have been chosen by a computer as the ideal match for Kammerer: she sang his songs at concerts, and she worked as his part-time assistant at the Vivarium. When she came home she lectured Lacerta about 'measuring rats' tails and taking their temperatures'. But she also gave Lacerta sound advice how to dress: 'Never wear pink or yellow after thirty. They make the young look young and the not-so-young, old. Don't wear flamboyant clothes; they are for the ugly. They divert people's eyes from the face to the clothes. You have no need for that.'

The young couple was socially much in demand. The future seemed to be laid out for them under a cloudless sky. The first innocent-looking cloud appeared in 1910 when Kammerer was thirty and received a letter from the great William Bateson at Cambridge, asking for 'a loan of a specimen of the midwife toad, showing the nuptial pads'.[1]

2

The campaign which was to embitter the rest of Kammerer's life was started, not as one would expect, by his colleagues at home, but in England; and its battles were fought not in German journals but in the austere columns of *Nature*. In Vienna itself Kammerer enjoyed great popularity not only among the artistic and social élite but also among his more imaginative colleagues. Among these were Professor Steinach, initiator of the rejuvenation treatment already mentioned; Professor Richard Semon, author of the classic treatise on *The Mneme*; Professor Hans Przibram and others. Steinach, Semon and Przibram gave him unqualified support in their scientific publications, and this carried considerable weight. So did Kammerer's uncontroversial textbook *Allgemeine Biologie— General Biology*—considered at the time as a standard work. The negative attitude of the University establishment was not

* Professor Przibram's daughter, now Countess Teleki.

so much directed against his Lamarckian theories, but rather an expression of the usual hostility of the grey birds in the groves of Academe against the coloured bird with the too-melodious voice. Nevertheless in 1918, at the age of thirty-eight, he had a good chance of being given an associate professorship (Freud got his only at the age of forty-six, and became a full professor at sixty-four). This happened at a time when he had finished writing *Das Gesetz der Serie*, a speculative hypothesis concerning the nature of meaningful coincidences (see Appendix 1). Przibram and other friends implored him to postpone publication until after the meeting of the University Senate which was to decide on his appointment. In keeping with his temperament, Kammerer refused the compromise.[2] That was the end of his hopes for a professorship—but it carried no reflection on his experimental work.

In Germany, too, his reputation remained high to the very end. To have a paper published in the *Archiv für Entwicklungsmechanik* was a great distinction for any biologist, and between 1904 and 1922 Kammerer was given more space in the pages of the *Archiv* than almost any other living scientist. His most serious opponent in Germany was the botanist Erwin Baur, but Baur's criticism was mainly directed at Kammerer's theoretical interpretation of his experimental work; it did not cast doubt on the work itself, or on his personal integrity. Though acid in style, it was a scientific, not a personal, dispute, which did no harm to the reputation of either side.

Thus, paradoxically perhaps, Kammerer's most implacable enemies were not the proverbially rabid *Herren Professoren*, but his reputedly good-humoured colleagues in a country for which he and brother Charley had such a naive admiration. The reality was different; it is difficult to convey the atmosphere of spite and unfair play which prevailed among students of evolution in the early years of the century. The man who had suffered perhaps more than anybody else from it was William Bateson. As he was the chief protagonist in the ensuing drama, we must try to understand the motives behind his actions.

3

Bateson had started as a Lamarckian. In 1886, when he was twenty-six years of age, he set out on an expedition to the Central Asian lakes (Lake Balkhash, Lake Tchalka, the Aral

Sea, etc.), hoping to find in the fauna of these unexplored and isolated waters evidence for the inheritance of acquired characteristics. He found none. His widow wrote in her preface to his book on the expedition:[3] 'His high hopes of successful elucidation of the problems he had set himself were disappointed; reluctantly, one by one, he put them aside.' He had spent some eighteen months in the Karakum desert, living with the natives, enduring severe hardships and repeated illness— and all for nothing, with no results to show on his return. It must have been a traumatic experience. I asked his son, Gregory Bateson, what kind of a person he had been. He replied:

> As to whether my father was a nice man, in spite of natural ambivalence, I incline to say yes. He certainly was liked by many people who were undoubtedly nice people. But notably those who liked him were in general not his colleagues. Many of the latter hated his guts.
>
> He was certainly not a nice man whenever the inheritance of acquired characteristics was mentioned. When this happened the coffee cups rattled on the table. Remember that he went to the [Siberian] steppe in order to prove the inheritance of acquired characteristics, using as data the creatures which he was going to find in Lake Balkhash, etc. The project was a complete failure. Perhaps this had something to do with his later attitude. I think he always knew that there was something very wrong with orthodox Darwinian theory, but at the same time he regarded Lamarckism as a tabooed pot of jam to which he was not allowed to reach. I have his copy of *The Origin of Species*, sixth edition, in which he listed on the fly-sheet the pages on which Darwin slipped into Lamarckian heresy.[4]

What drew him originally to the tabooed pot of jam was the crisis which Darwinism underwent during the last decades of the previous century. 'In the study of evolution,' Bateson wrote in retrospect, 'progress had well-nigh stopped. The more vigorous, perhaps the more prudent, had left this field of science to labour in others, where the harvest is less precarious or the yield more immediate. Of those who remained some still struggled to push forward the truth through the jungle of phenomena: most were content supinely to rest on the great clearing Darwin made long since.'[5]

In fact the jungle had become impenetrable even in Darwin's lifetime. In 1867 a Professor of Engineering at Edinburgh University, Fleeming Jenkin, published a review of *The Origin of Species* in which he demonstrated, by an astonishingly simple logical deduction, *that no new species could ever arise from chance variations by the accepted mechanism of inheritance.* That mechanism, known as the 'blending of inheritance', was based on the apparently commonsensical assumption that the native equipment of the new-born babe was a mixture or blend of the characteristics of the parents, to which mixture each parent contributed approximately one half. Francis Galton, Darwin's cousin, called this 'the law of ancestral inheritance', and gave it a mathematical formulation. Assuming now that an individual with a useful chance variation cropped up within the population and mated with a normal partner (that is, excluding the very unlikely case that it met and mated with a partner possessing the same chance variation), then its offspring would inherit only fifty per cent of the useful new feature, its grandchildren twenty-five per cent, its great-grandchildren twelve and a half per cent, and so on, until the useful novelty vanished like a drop in the ocean, long before natural selection had a chance to make it spread.

It was this objection which so shook Darwin that he inserted a new chapter into the sixth edition of *The Origin of Species*, resuscitating the inheritance of acquired characters. As his letters to Wallace clearly indicate, he saw no other way out. His son, Francis Darwin, later commented: 'It is not a little remarkable that the criticisms, which my father, as I believe, felt to be the most valuable ever made on his views, should have come, not from a professed naturalist but from a Professor of Engineering, Mr. Fleeming Jenkin.' It is no less remarkable, as Sir Alister Hardy noted, that 'the great brains of the Victorian era' should not have realised the basic logical fallacy which Jenkin had pointed out.

Bateson, however, did realise it, and also other weaknesses in the Darwinian edifice. Hence the despairing reference to the 'jungle of phenomena'; hence, also, his early hope, based on the same reasoning as Darwin's, to find evidence in the Asian lakes for a Lamarckian mechanism of inheritance. That hope was cruelly frustrated—and its echo reverberated only in the rattling of the coffee-cups. A few years later, however, by an unexpected and almost melodramatic turn of events, the crisis was resolved, the clouds vanished, and Darwinism became transformed into neo-Darwinism.

The crucial event was the rediscovery, in 1900, of a paper read thirty-five years earlier by an Augustinian monk, Gregor Mendel, to the Natural History Society of Brünn in Moravia. The paper, called 'Experiments in Plant Hybridisation', had been published in 1865 in the Proceedings of that Society, and completely ignored by the rest of the world. Mendel died nineteen years later, a sadly disappointed man, though he might have found some consolation in becoming the Abbot of his monastery. Another sixteen years later his paper was rediscovered, almost simultaneously and independently, by three biologists in three different countries: Tschermak in Vienna, de Vries in Leyden, Correns in Berlin. Each had been searching the literature for a clue to indicate a way out of the cul-de-sac, and each saw immediately the revolutionary significance of Mendel's hybrid garden peas—which, like Newton's apple, were to become an integral part of science-lore. Mendel's plants showed that the units of heredity—later to be called genes—did not 'blend' and thus become diluted; they were rather like hard, stable marbles which combined into a great variety of mosaic patterns, but preserved their identity and were transmitted unchanged to the next generation, to combine into new kaleidoscopic patterns and become reshuffled again at the various stages of the reproductive process. If a mutation occurred—a random variation in a gene—and it had survival value, then it was not whittled away in successive blendings, but preserved by natural selection. Now everything was falling into place. Every unit of heredity was incorporated in a Mendelian gene, and every gene had its allotted place on the chromosomes in the cell-nucleus, like beads on a string. Evolution no longer had any secrets. Or so it seemed—for the time being.

The pioneer of Mendelism in England was William Bateson. In the year 1900, when he was forty, he read Mendel's paper. He read it in the train on his way to deliver a lecture to the Royal Horticultural Society. He rewrote his lecture then and there in the light of the new revelation. Its impact on him is reflected in the preface to his book on *Mendel's Principles of Heredity: A Defence*, published in 1902. I have quoted from it the passage about progress in evolution having 'well-nigh stopped' on the clearing in the jungle that Darwin made. The preface continues:

Such was our state when two years ago it was suddenly discovered that an unknown man, Johann Gregor Mendel, had, alone and unheeded, broken off from the rest. This is no metaphor, it is simple fact. Each of us who now looks at his own patch of work sees Mendel's clue running through it: whither the clue will lead we dare not yet surmise.'

It is the same language of jubilant enthusiasm in which Kammerer announced the Brave New World where acquired learning became inheritable. This enthusiasm for theories was a quality that both men shared, however different their characters in other respects. It affected not only their professional outlook, but also their private lives. Kammerer called his daughter Lacerta, after his favourite lizards. Bateson called his youngest son Gregory, after his hero Gregor Mendel.

'Whither the clue will lead.' It led, predictably, into trouble. Bateson was the first to demonstrate by his experiments with poultry that Mendel's laws of inheritance applied to animals as they applied to plants. But all the evidence in the world could not convince the influential group of anti-Mendelians who clung to the 'blending' theory of inheritance and tried to prove it by a statistical approach to biology. They called their method Biometrics—it had been originated by Francis Galton; and they published a journal, *Biometrika*, edited by the mathematician Karl Pearson and W. F. R. Wheldon. The controversy went on for several years, and was conducted—particularly on the side of the biometricians—with a remarkable degree of bitchiness. Pearson and Wheldon attacked Bateson and other Mendelians, but did not allow them to reply in the columns of *Biometrika*. Bateson retaliated by adopting the strategy of the Trojan Horse: he had his reply privately printed and bound in a cover which was an exact replica of the cover of *Biometrika*. Pearson used his considerable influence with the editors of *Nature* to prevent publication of Bateson's letters. Among the papers Bateson left* there is a letter from the editor of *Nature*, dated May 19, 1903; the editor was 'not prepared to continue the discussion on *Mendel's Principles* and therefore returns herewith the papers recently sent to him by Mr. Bateson'. A remarkable document,

* The 'Bateson Papers' (including correspondence and notes) have been deposited by Professor Gregory Bateson in the Library of the American Philosophical Society; I am indebted to him for granting me access to them. I shall henceforth refer to them as 'the Bateson Papers'.

considering that at next year's Meeting of the British Association at Cambridge, Bateson was President of the Zoology section and 'delivered a stirring address vindicating the methods of Mendelian research, and challenging the conceptions of the Biometrical school. The heated debate which succeeded it was keenly followed by a crowded audience, and at the end of it there was no mistaking the feeling that the exponents of Mendel had made good.' This inspired passage is quoted from Bateson's obituary in *Nature*—which, of course, has always been on the side of the angels.

4

Such, then, was the lovely atmosphere in Cambridge, in the years preceding the Bateson-Kammerer dispute. But there were two profoundly ironical twists in the story of the rise of neo-Darwinism, which are little known, and which I must briefly relate.

First of all, Gregor Mendel's statistics in that classic paper were faked—or, to use a more polite term, doctored. Bateson did not know this—but had he known, one wonders whether it would have made much difference to his attitude; whether it would have changed his conviction that the theory was right; and if not, whether he would have broadcast his knowledge, or kept it prudently to himself in the interest of higher truth. Be that as it may, the doctoring of Mendel's results was only discovered in 1936, by which time Bateson, Pearson and the other actors in the drama were dead. Mendel's laws of inheritance postulate that if you crossbreed two varieties of a species—say, tall plants and dwarf plants—then in the offspring of the hybrids, the ratio of tall to dwarf plants will *approximately* be as 3:1.* Mendel's published experimental results gave the actual ratio so close to the expected 3:1 that it was far too good to be true; never has any of the thousands of researchers engaged in this type of experimental work produced ratios so close to the ideal 3:1 value. It was the late Sir Ronald Fisher, the greatest statistical mathematician of his time, who proved that the detailed figures in Mendel's paper could not have been true

---

* The gene for tallness, 'T', is dominant, for dwarfness, 'd', recessive; hence in the four possible combinations: TT, dT, Td, dd, the first three will produce tall plants and only the last one dwarfs.

—because it was inconceivable, short of an 'absolute miracle of chance', to obtain these ratios.*

It is rare to find this historical scandal mentioned in the literature. It was not so much hushed up as shrugged off. Since Mendel's Laws had been shown to be correct, what does it matter if he cheated a little? Characteristic of this tolerant attitude is the following comment by Sir Alister Hardy in his Gifford Lectures:

I do not think that anyone supposes that Mendel himself deliberately faked the results. It seems most likely, and this, as Fisher suggests, perhaps makes Mendel an even greater figure than one hitherto thought, that, instead of his first doing a vast number of experiments and *then* coming to his conclusion, he, after a few trials, worked out his theory by mathematics in his study and then put it to the test. One can imagine him telling his gardeners of his theory, why he was doing the experiments, and why he expected to get a ratio close to, but not exactly, 3:1. As the experiments proceeded, the gardeners, who helped him, no doubt saw quite clearly that the results were coming out as he had foretold; and, assisting in the counting, it must be supposed that they saved themselves much trouble by giving Mendel the results as he had foretold, not exactly but very nearly 3:1—too near as it turned out! With painstaking care Mendel himself no doubt carried out the pollinations but perhaps left the mere counting to his assistants. We shall never know the exact truth of this story, but it appears to stand . . . as another example of brilliant insight being confirmed by experiment.[6]

One is tempted to smile at the generous use of the expressions 'most likely', 'no doubt', 'one can imagine', 'it must be supposed', 'perhaps', etc. Tolerance and broadmindedness towards the dead are no doubt laudable; but what if that obscure monk in Brünn had been caught red-handed doctoring his statistics—or even neglecting to check the gardeners' statements?

Another puzzling aspect of the affair is why Karl Pearson and the other biometricians, who specialised in statistical mathematics and who were anti-Mendelians, did not discover the flaw in

* The ratio of the four kinds of possible gene combinations by pollination is governed by the laws of probability, as in a game of dice, and cannot be influenced by the experimenter's skill.

Mendel's figures—which, once they have been pointed out, are quite obvious to anybody with an inkling of the mathematics of probability. In the heat of a controversy concentrated on one particular aspect of a problem, scientists are apt to behave as if they were wearing blinkers, just as ordinary mortals.

And now to the second ironic twist in this story. As we have seen, for thirty-five years, between 1865 and 1900, nobody paid any attention to Mendel—with one exception. The German botanist Wilhelm Focke published in 1881 a book on plant-hybridisation, *Die Pflanzen Mischlinge*, which contained a passing reference to Mendel's experiments; and thus Mendel's name found its way into the bibliographical references of this one book. Now it so happened that just at the time when it was published, the Oxford biologist, George John Romanes (the founder of the Biannual Romanes Lectures), was engaged in writing the article on hybridisation for the ninth edition of the *Encyclopædia Britannica*. When he had finished it he felt the need to append an impressive bibliographical list. So he wrote (on November 30, 1880) to Charles Darwin and asked him for a list of suitable works on the subject. Darwin, equally bent on avoiding unnecessary trouble, sent him Focke's book *Pflanzen Mischlinge*, which had just arrived, explaining that he had not found the time to read it, but that it had a nice bibliography which would just suit Romanes' purpose. It did. Romanes simply copied out the bibliographical references, in the same order as in *Pflanzen Mischlinge*, regardless whether or not they were referred to in the text of his article—a procedure decidedly odd. Thus Mendel's famous paper was actually mentioned in the ninth edition of the *Encyclopædia Britannica*—in the bibliography, but not in the text of Romanes' article. In this modest hiding place it appeared while both Darwin and Mendel were still alive, and stayed there coyly for twenty years, without attracting any notice, until it suddenly exploded like a time-bomb.

Some years ago there was a considerable scandal when an English journalist published a biography of a famous European and was convicted of having lifted his bibliographical list from a foreign work. Surely, one would have thought, such deplorable practices may happen among journalists, but never in the austere world of academic science and the *Encyclopædia Britannica*.

57

I have dwelt at some length on the mental climate and polemical methods of the period, in order to show that the climate was far from moderate, and the methods far from scrupulous. It is only against this background that the campaign against Kammerer can be seen in proper perspective.

Five

The first skirmish between Bateson and Kammerer took place in 1910 when Bateson was fifty, Kammerer thirty, and both were eminent in their respective countries and fields. Bateson read Kammerer's early papers on the nuptial pads of the midwife toad and the coffee-cups started to rattle. 'We believe such things when we must, but not before', he wrote in *Problems of Genetics*[1]—the book on which he was working at the time, but which was only published three years later, in 1913. Chapter IX of the book was meant to put Lamarck's ghost to rest by a final refutation of the inheritance of acquired characters, 'a process frankly inconceivable'.[2] He had dealt earlier on with other Lamarckian heretics, among them Professor W. L. Tower of Chicago, whose work, Bateson wrote, 'though offered with every show of confidence, exhibits such elementary ignorance, both of the special subject and of chemistry in general, that it cannot be taken into serious consideration'.[3]

But no sooner was the dragon's head cut off when another grew in its place. And Kammerer's was a more formidable head than Tower's—as Bateson at once realised. He wrote (in *Problems of Genetics*):

> The series of experiments made by Kammerer with various amphibia have attracted much attention and have been acclaimed by Semon and other believers in the transmission of acquired characters as giving proof of the truth of their views. With respect to these observations the chief comment to be made is that they are as yet unconfirmed. Many of the results that are described, it is scarcely necessary to

say, will strike most readers as very improbable; but coming from a man of Dr. Kammerer's wide experience, and accepted as they are by Dr. Przibram, under whose auspices the work was done in the *Biologische Versuchsanstalt* at Vienna, the published accounts are worthy of the most respectful attention.[4]

Bateson goes on to say: 'I wrote to Kammerer in July, 1910, asking him for the loan of such a specimen [of *Alytes* showing the nuptial pads], and on visiting the *Biologische Versuchsanstalt* in September of the same year, I made the same request, but hitherto none has been produced.'[5] To this sentence Bateson affixed the following footnote: 'In reply to my letter Dr. Kammerer, who was then away from home, very kindly replied that he was not quite sure whether he had killed specimens of *Alytes* with "Brunftschwielen" [nuptial pads] or whether he only had living males of the fourth generation, but he would send illustrative material.'

This accusation—that Kammerer did not, in 1910, send a specimen of *Alytes* to Cambridge, and did not show one to Bateson when he visited Vienna later in the same year—was to play an important, and indeed decisive, part in the controversy. The exact wording of Bateson's request and Kammerer's reply is therefore worth recording.*

17.7.1910
Herrn Dr. P. Kammerer
Dear Sir,
    I have been reading with interest your various papers on inheritance of the effects of conditions with a view to writing a chapter on the subject in a book which I have in hand. In common with others who are working at Genetics here it would very greatly interest me to see specimens illustrating the changes produced.
    In most cases the alterations are so much matters of degree and age, or stage of development, that I can well understand that it is not easy to demonstrate the effects to those who have not witnessed the course of the experiment.
    In one instance however that of the development of *Brunftschwielen* on all the males of the 4th generation

* The Bateson Papers. Bateson's letter to Kammerer is preserved in a handwritten copy; Kammerer's reply in the original. (It was written in English.)

60

(p. 516) of the *Alytes obstetricans* treated it is easy to demonstrate the change produced. I venture therefore to ask if you would be so good as to lend a specimen for a short time showing the development. Naturally I will undertake to examine it with the greatest care and to return it uninjured immediately or with the shortest possible delay.

Trusting that you may see your way to grant this favour,

I am

Yours faithfully

William Bateson

Steinbach am Attersee
Gasthof Föttinger
Ober-Österreich
22.7.1910
Dear Sir,

I am just enjoying my holidays, and so your kind letter has been sent from Vienna to Steinbach. Therefore my answer comes late.

As soon as I shall be returned to my usual work—two congresses and a journey to Munich are still between—I will send to you any objects you may need for your book and have interest for, with the greatest pleasure! I hope, that it will not be too late then for using them in the chapter 'Effects of external conditions' of your future book.

I am not quite sure, whether I killed already specimens of *Alytes* with 'Brunftschwielen', or am possessing only living males of this (F_4) Generation.

But I do not doubt that also other objects are well fitted to show easily the effect of conditions and their inheritance. Especially my new experiments on influence of soil etc. upon colours (not yet published, except some preliminary notes, for instance in the 'Verhandlungen Deutscher Naturforscher u. Ärzte', Salzburg 1909) are much more favourable for that purpose than the instinct variations, in spite of their morphological consequences.

I have also promised (i.e. Dr. Przibram has in my name) to Mr. Doncaster, to spare him a series of tadpoles with alterations etc. for your museum; and it is my intention, to fulfil this promise together with that given to you in my present letter, during the beginning of this autumn.

Yours faithfully

Paul Kammerer

A careful reading of Kammerer's reply shows that he did *not* promise to send Bateson an *Alytes* showing the nuptial pads because he was not sure whether he had any preparation showing them. What he offered was to send 'other objects', which he thought were 'much more favourable for that purpose'— namely, to demonstrate the inheritance of acquired characters. Bateson's carefully worded letter obviously made Kammerer believe that the 'chapter on the subject in a book which I have in hand' was to be dedicated to that purpose.

Bateson's purpose was of course the exact opposite. The chapter in question (Chapter IX) in *Problems of Genetics* has twenty-six pages, thirteen of which were devoted to a scathing criticism of Kammerer's experiments. But Kammerer saw Bateson's book only much later.

2

This exchange of letters between Bateson and Kammerer took place in July, 1910. Two months later Bateson visited Vienna, and that visit changed his attitude to Kammerer from scepticism to hostility, which later verged on obsession. Bateson was a guest in Przibram's luxurious flat on the Parkring (the same where Przibram senior had installed electric lighting); and he disliked everything he saw. In a letter to his wife he wrote:[6]

28.IX.1910. Parkring, 18. Wien.
A very strenuous day! It seems I am not to be allowed to do anything *ohne Begleitung*, a rather severe penalty for the comfort I am otherwise to enjoy. Przibram took me to the Kunst-historische Museum in the morning, but somehow I didn't feel to get as much as I might have done. His artistic sympathies are rather narrow, and his thoughts run in prosy lines. I am to have a most commodious flat, containing all manner of solid comforts. The valet gets in almost as many 'Herr Professors' into his speech as definite articles. He has unpacked my clothes and still finds me too grand to be put into the compass of a plain 'Sie'. I wanted to get to *Rigoletto* tonight, but Przibram declares for a quiet evening and such it is to be. His wife is tired and has not yet appeared. He and I are to dine at a Restaurant—we have been talking all day and I begin to feel that our resources are exhausted. Tomorrow morning he is happily

engaged at the Physiological Congress, which I mean to avoid.

I had a long spell with Kammerer—and there is no denying any longer the extraordinary interest of what he is doing. The *Brunftschwielen* [nuptial pads] cannot be produced. Somehow or other I have hit on a weak spot there. This criticism must have been quite unexpected, and they don't seem to have noticed how critical the point must in any case be. However they say that in course of time they will get more specimens, &c. But he has certainly done a very fine lot of things, and he comes uncommonly near showing that an acquired adaptation is transmitted. I don't like it, and shall not give in till no doubt remains.

Kammerer is not an ordinary man. There is something inclining to the artistic about him, and I understand at one time he thought of being a musician. I hardly like to say so seriously, but there was just a faintest tinge of something like suspicion of humbug in my mind once or twice, but taking the whole series of experiments together I cannot really entertain the idea of fraud. It would have to be deliberate fraud on a large scale, pursued for years, to produce his things. His face reminded me a little of Pearson's, and I think it may have been that which raised the momentary doubt of his candour.

29.IX.10. Well! A deadly dull evening we did have. I don't find Przibram a sympathetic companion. I wish I had a little more liberty, and doubt very much whether this bondage will even be cheaper in the end. All these 'Küss die Hands' and 'Herr Professors' may not improbably eventuate in tips. . . .'

There are several revealing points in these letters. The first is the resemblance—real or imaginary—to Karl Pearson, Bateson's mortal enemy in the days of the *Biometrika* controversy. One does not have to wade into deep psychology to realise that associations by physical likeness can give rise to strong and lasting emotional reactions—infatuation or detestation, as the case may be.

The second point is the grudging admiration for Kammerer ('not an ordinary man'—'has certainly done a fine lot of things' —'I don't like it, and shall not give in', etc.). Bateson set out for Lake Balkash to prove the inheritance of 'acquired adaptations' and was unable to do so. Kammerer came 'uncommonly near to proving it'—without spending eighteen months with the savages in the steppes and feeling like a fool afterwards. Was he to succeed where Bateson had failed?

The most revealing passage, however, the significance of which only dawns on one in the light of what has been said before, is the remark 'The *Brunftschwielen* cannot be produced. *Somehow or other I have hit on a weak spot there.* This criticism must have been quite unexpected.'

In other words, Bateson thought that he had found Kammerer's Achilles heel. He had no doubt been shown around the aquarium and terrarium, populated with Kammerer's salamanders, lizards, long-siphoned *Ciona*, and the rest of the 'fine lot of things'—but he hadn't been shown the nuptial pads, because there were none to show outside the mating season on live specimens, and Kammerer had not killed any of his precious breeders when they did show them. Hence Bateson's criticism was 'quite unexpected' and was no doubt received by Kammerer and Przibram with mild surprise; they were unable to understand 'how critical the point in any case must be' since Kammerer himself had never claimed the pads to be a decisive argument.

In *Problems of Genetics* Bateson also criticised, besides the *Alytes* experiments, Kammerer's work with salamanders. In the same year, 1913, in which the book appeared, Kammerer published in the *Archiv* his—up to then—final report on the colour changes in the salamander experiments, which made Bateson's objections outdated. But Bateson never revised his criticisms, nor did he return to the *Salamandra*. By then his strategy was set: to concentrate on the midwife toad, and ignore the rest of Kammerer's work.

Six

During the war, communications were, of course, interrupted. In 1919 Kammerer published in the *Archiv für Entwicklungsmechanik* his hitherto most detailed paper on the midwife toad. It also contained his answer to several critics, including Bateson. I shall quote its relevant passage at length; but first, let me remind the reader that to make that land-mating toad *Alytes* mate in water, and to breed further generations from the waterlogged eggs deprived of the male's care, was an experiment which nobody else had been able to repeat:*

> To send precious specimens abroad is not always easy or advisable. I shall not speak of the sad experiences I occasionally met when lending out specimens; this is the obvious reason why most Museums and Collections made it a rule to allow nothing to be taken outside the building. Nevertheless I intended to send a pad-equipped toad to England. The shortage of appropriate specimens, which only permit one just to scrape along from one generation to the next, made me hesitate each time. It should be remembered that quite often the continuation of a genetic line depends on a single specimen; by no means all are willing to mate [under the artificial laboratory conditions] even when they develop the morphological attributes of the mating season. And it is impossible to tell beforehand whether a given individual will continue to breed or whether it could be sacrificed as

* See Appendix 3 on the controversy with Boulenger, who tried but failed.

useless. . . . [Follow references to photographs in the text and of specimens on exhibit in the Vienna Institute.]

I would like to take this opportunity to make a few remarks about the preservation of specimens, a problem which has worried me for a long time. On most occasions, one is confronted with the choice: either to kill the specimen and preserve it at the risk of having to abandon the continuation of the experiment (that, for instance, happened in the case of the large-eyed specimens of *Proteus*!), or to keep them all alive as far as possible, at the risk of being finally left with corpses which are no longer fit to be preserved. Particularly in the earlier years, when I met with less scepticism, I was reluctant to kill an experimental animal, and kept trying to achieve, by the breeding of further generations, even more clearcut results. This method alone—though from certain points of view open to objection—enabled me to achieve success in many breeding experiments when the outcome was touch and go—that is, entirely dependent on just one particularly virile specimen. Against this ultimate dependence on a single individual there is no certain protection, even if you start with the largest possible number of experimental animals—as I always did.

The demand to furnish exhibits has been more often and more peremptorily addressed to me than to other researchers. When Driesch [leading embryologist of his time] visited our Institute and we asked him for a donation of exhibits for our Collection, he replied, 'I have preserved nothing; anybody who doubts my experiments should repeat them!' That somebody should at long last repeat mine is my most fervent wish.

Bateson, too, visited our Institute at the beginning of September, 1910, after the International Zoological Congress at Graz. He remarks on page 202 [of *Problems of Genetics*]: why did I not show him at that time *Alytes* with nuptial pads? Well, for the reasons just explained I had no dead preparations; and in living specimens the pad appears, as generally known, only periodically during the mating season when the animal is on heat. But at the time [of Bateson's visit] the mating season was still ahead; a few weeks later the black discolouration and the growth of the horny papillae would already have been visible. And of course there is no shortage of specialists who have seen the pad at the proper period.[1]

Later on,[2] Kammerer wrote with some bitterness that he did not continue to work with *Proteus*, to see whether the large eyes became hereditary, because the critics would have found fault with the experiment anyway. He decided instead to kill off his specimens, and it was thanks to this circumstance that these were preserved as evidence. Had he not killed off his breeders, his claim that *Proteus* may be induced to develop a normal eye would no doubt have met with the same scepticism as the nuptial pads of *Alytes*.

2

Contacts between the scientific communities of the former belligerents were only resumed slowly. The first to call the attention of English scientists to Kammerer's new paper was E. W. MacBride, Professor of Zoology at the Imperial College of Science, London. MacBride has been described by one of his adversaries, Professor Darlington, as 'a stout little Irishman with a shrill voice and, like his compatriot G. B. Shaw, a pugnacious Lamarckian'.[3] He was also a devoted Kammererian. In his letter, which *Nature* published on May 22, 1919, he gave a brief summary of Kammerer's experiments, then quoted some of Bateson's criticisms and Kammerer's reply to Bateson in the *Archiv für Entwicklungsmechanik*. He concluded:

> It must, we think, be conceded that Kammerer has fairly taken up the gauntlet thrown down to him by Professor Bateson and the present position of the matter is that a strong *prima facie* case for the inheritability of acquired variations has been made out. Doubting Thomases could be convinced only by a journey to Vienna and an inspection of the modified males, for it is unreasonable to expect Kammerer to send these priceless specimens to any zoologist who chooses to doubt his word. It is to be hoped that, once peace is signed, this journey will not be delayed.
>
> It may perhaps be said that no notice should be taken of Kammerer's results until some other investigator repeats them. Such a course is not pursued with regard to any other zoological investigations. When new discoveries are published, we thankfully receive them. We keep perhaps an

open mind until they are repeated, but freely concede that a *prima facie* case has been made out for them.

To Mendelian critics I would point out that the difficulty of instituting experiments designed to test the inheritability of acquired characters is colossal. I have persuaded Mr. E. G. Boulenger,* Curator of Reptiles, to make preliminary arrangements to have some of Kammerer's experiments repeated in the Zoological Gardens. I found that a minimum of six years would be required before decisive results could be obtained. This new paper of Kammerer's appears to represent the result of seven or eight years' work. The proper rejoinder of the Mendelian is not to jibe at the absence of confirmatory evidence from other investigators (and some even of this is available), but to obey the Scriptural injunction, 'Go thou and do likewise'.

Bateson's reply to MacBride (*Nature*, July 3, 1919) was moderate in tone, rich in its veiled insinuations. He dismissed Kammerer's fifteen years of work with *Salamandra* with the single remark that 'salamanders corresponding with Dr. Kammerer's several patterns can be had from the dealers'. It will be remembered that Kammerer claimed to have produced a hereditary transformation of the black salamander into the spotted variety; the implication of Bateson's remark was that Kammerer had simply bought his spotted specimens 'from the dealers'. Six years earlier, in *Problems of Genetics*, Bateson's 'chief comment' on Kammerer's experiments had been 'that they are as yet unconfirmed'; yet 'worthy of the most respectful attention'. Now Bateson implied that the results had been obtained by fraud.

A little reflection would have shown that the substitution of a specimen bought from the dealers for an experimental animal was a technical impossibility because the appearance and growth of the yellow spots and stripes in the originally black animals was a gradual process spread over six years; and the staff of the Viennese Institute, headed by Przibram, whose integrity nobody questioned, could not have failed to notice the substitution in Kammerer's aquariums and terrariums exposed to public view. Had Bateson taken the trouble to read Kammerer's 1913 publication† and looked at the charts and illustrations in

* The younger Boulenger.
† See above, p. 64.

68

its 189 pages, he would have realised the untenability of his suspicion.

Bateson's letter to *Nature* then recapitulated the 1910 episode. In *Problems of Genetics* he had merely mentioned, in a footnote, that Kammerer, being away from home, 'had very kindly replied' that he was not sure whether he had killed any *Alytes* with pads. Now Bateson felt moved to quote Kammerer's 1910 letter in full (see p. 61). What the letter shows is that Kammerer had enthusiastically and somewhat rashly promised to 'send you any objects [*other than Alytes*] you may need for your book and have interest for, with the greatest pleasure!' Later in that summer Bateson had visited Vienna, as described above, and after that there was no more question of sending specimens.

The dates of the 1910 exchange of letters are significant. Bateson wrote from Cambridge to Vienna on July 17; Kammerer was on holiday, and the letter had to be forwarded, yet he replied on July 22; there was no airmail in those days, and the Viennese are notoriously slow in answering letters. So Kammerer must have been very keen to co-operate. Yet the obvious purpose of Bateson's quoting Kammerer's letter in full, after nine years, was to give the impression that Kammerer had made a promise and failed to keep it.

The rest of Bateson's July, 1919, letter to *Nature* was devoted to the nuptial pads. He criticised the photographs of the pads in Kammerer's latest paper; expressed doubts whether the pads were 'in the right place' on the toad's hands, and concluded: 'Professor MacBride urges that sceptics should repeat experiments on the inheritance of acquired characters. We, however, are likely to leave that task to those who regard it as a promising line of inquiry. Even in this case of *Alytes*, were a male with incontrovertible *Brunftschwielen* before our eyes, though confidence in Dr. Kammerer's statements would be greatly strengthened, the question of interpretation would remain.'

Ten years had gone by since Bateson had first heard of the midwife toad, and now, approaching sixty, his attitude had hardened. Ten years earlier he wrote that if the existence of the pads were confirmed, they 'would go far to proving Kammerer's case'. Now, confronted with the new evidence in Kammerer's 1919 paper, he took the line that even if they existed, they would not prove anything ('the question of interpretation would remain').

We remember that Kammerer himself did not regard them as proof of the inheritance of acquired characters; to him the *Salamandra* and *Ciona* experiments were far more important. The battle of the nuptial pads was not of his own choosing; but he was forced into it.

3

Bateson, thinking he had found Kammerer's 'weak spot', pursued it relentlessly. On July 20, 1920, he wrote to a zoologist, Mr. Martin, who was preparing to visit Vienna:

> In Vienna call of course on Dr. Hans Przibram (pronounced Pr̆schi-brahm) of the *Biologische Versuchsanstalt.* We are in correspondence and he will be glad to see you. Both he and his wife are capital people. At his Laboratory you will see the famous *Alytes* of Kammerer's making. If you do see it note very carefully the position of the alleged *Brunftschwielen* on EACH side. Make them take it out of the bottle. Chr. Bonnevie saw it in the bottle but could make nothing out. Bear in mind that a good deal of grafting is nowadays possible—but I would not suggest that overtly [*sic*]. Make sure too that the animal *is* an *Alytes* (a little thin toad with small, weak hands) but I can't give you a certain diagnostic. Przibram will have Boulenger's *Batrachia of Europe* with good figures. Ask to have the specimen out in a dish where you can *examine it quietly with a dissecting lens at your leisure* (Italics in the original).*

Neither Kammerer nor Przibram seemed to have realised the degree of Bateson's hostility, and to what lengths he was prepared to go. Przibram, who thought that Bateson was still his friend, kept writing affectionate letters to him. Kammerer himself, in the summer of 1920, sent Bateson microtome slides of the nuptial pads. Bateson did not acknowledge them. But on September 6, in reply to a letter from Przibram, he wrote:

> Some weeks ago I received two slides sent, no doubt, by Dr. Kammerer—the one labelled as normal, the other showing the structure claimed to be produced by treatment.

* In a handwritten note appended to the copy of this letter Bateson's widow wrote: 'I cannot remember that Martin saw the specimen. I fancy the Przibrams were not in Vienna when he was there and that he saw nothing of K's experiments.'[4]

Without knowing more than I yet do as to the cycle of events in normal *Alytes* I can form no opinion as to the significance of these specimens; and beyond expressing my thanks for the kindness which led him to send these slides I can say no more at present. P.S. You will let me know if Dr. Kammerer would like to have the slides returned soon. There are some naturalists in England who would like to see them and I have not hitherto had an opportunity of showing them to them.[5]

Przibram replied (on a picture postcard with an idyllic landscape, painted by himself):

Vienna, Sept. 17th
Dear Professor Bateson, I have received your letter dated September 6th with thanks and beg you to keep the slides of Alytes as long as you like. We have others for our Museum and if you think it worth keeping the specimens sent altogether or giving them to an English collection where they may be appreciated it will be a pleasure to us.

Yours truly,
H. Przibram.[6]

Bateson kept the slides, but did not mention them in his attacks on Kammerer—except to suggest that Kammerer had substituted sections from some other toad or frog normally endowed with pads and pretended they came from *Alytes*. He made the same suggestion to E. G. Boulenger (the son); however, even Boulenger baulked at it. After a visit to the Vienna Institute in 1922, he wrote to Bateson:

I had a long talk with Przibram who not only takes full responsibility for all K's experiments, but states that he was present when the sections were cut which prove that the black colour is due to horny substance, and not pigment. If we still disbelieve we must assume that Przibram is a dishonest person.[7]

The implication was clearly that such an assumption was absurd and that the slide must be accepted as evidence.

Earlier in the same year, J. H. Quastel—at that time a graduate student at Trinity College, Cambridge, later Professor of Biochemistry at McGill University—visited Vienna with the assignment to interview Kammerer and supervise the taking of photographs of his specimens. The Royal Society report of the Ordinary Meeting on January 18, 1923, contains the following communication:

Prof. E. W. MacBride, F.R.S. Remarks on the Inheritance of Acquired Characters. (Verbal communication only.)

It is well known to zoologists that during the last fifteen or twenty years a series of experiments have been carried out by Dr. Paul Kammerer at Vienna, which tend to show that acquired qualities or, in other words, modifications of structure induced by modified habits, are inheritable. The results of these experiments have been received with much scepticism, both here and on the Continent, and the *bona fides* of Dr. Kammerer has been called in question.

One of the most interesting of his experiments was in inducing *Alytes*, a toad which normally breeds on land, to breed in water. As a result, after two generations, the male *Alytes* developed a horny pad on the hand, to enable him to grasp his slippery partner.

It has been admitted by Kammerer's critics that if he could demonstrate this pad, he would to a large extent succeed in establishing the validity of his results.

This summer at my request Mr. J. Quastel, of Trinity College, Cambridge, when in Vienna, interviewed Kammerer, and was shown by him one of these modified males. Quastel photographed the animal, and enlargements from his photographs are now shown. Subsequently, at my request, the Zoological Society despatched Mr. E. G. Boulenger on a visit to Vienna. He, too, saw the modified male, and was assured by Przibram, the head of the Biological Institute, that all Kammerer's experiments had been done under his (Przibram's) supervision, and were perfectly genuine.

Seven

I

The decisive episode in the controversy was Kammerer's visit
to England in 1923, which brought about his second confronta-
tion with William Bateson.

The visit was sponsored by the Cambridge Natural History
Society, and financed by a group of enthusiastic undergraduates
—among them young Gregory Bateson. Kammerer lectured to
the Cambridge Society on April 30, and repeated the lecture
on May 10 to the Linnean Society in London. *Nature* published
the full text on May 12. He also brought with him a collection
of exhibits, including specimens of *Proteus, Salamandra* and—
most important to his opponents though not to him—the last
and only specimen of a midwife toad preserved in alcohol,
showing the nuptial pads on its right fore-limb, while on the
left fore-limb the pad had been excised and used for the preparation of sections which were shown on microscope slides.

It was the last and only specimen, because during the war
most of the experimental animals in the Vivarium had died, and
most preparations had perished. The laboratory assistants and
trained keepers had been called up to arms, and the Institute
was 'an utter wreck'.[1] Felicitas, Kammerer's devoted wife, tried
for a while to look after the delicate creatures, but they required
expert knowledge to be kept alive, and the plants controlling
temperature and humidity were out of action. The seventh
generation (F_6) of 'water-mating' *Alytes*, hatched in 1914,
developed oedemas on their fore- and hind-limbs, followed by
rapid ulceration, and the breed died out. A *single surviving male*

(of the F_5 generation), which had just entered puberty, was preserved, and 'provided the opportunity to observe the development of the thumb pads, although it never got to using its fore-limbs to grasp a female'.[2] This ten-year-old preparation, preserved in a glass jar, was the crucial exhibit—and subsequently became the main cause of Kammerer's suicide.

Kammerer himself during the war had been exempted from active military service owing to a heart condition, but he had been detailed to the military censorship department, which left him little time for anything else. Professor von Bertalanffy, now at New York State University, still remembers travelling with him on suburban trains in the early-morning darkness to town, Kammerer serenely reading by the light of a pocket torch.

After the war and the collapse of the Austro-Hungarian monarchy came political unrest and the catastrophic inflation which, within a few months, ruined the Central European middle classes, including the once-prosperous Kammerers and Przibrams. To keep going somehow, Kammerer had to concentrate more and more on lectures and the writing of popular articles.

Thus at forty-two, after the loss of his cherished salamanders, toads and lizards, Kammerer was rather in the position of a writer whose manuscripts have been lost in a fire. His Cambridge lecture ended on a sadly resigned note:

> The present circumstances are scarcely favourable for the furtherance of these researches in heredity in my impoverished country. During the War experimental animals, the pedigrees of which were known and had been followed for the previous fifteen years, were lost. I am no longer young enough to repeat for another fifteen years or more the experiments, with the results of which I have been long familiar. . . . The necessities of life have almost compelled me to abandon all hope of pursuing ever again my proper work—the work of experimental research. I hope and wish with all my heart that this hospitable land may offer opportunity to many workers to test what has already been achieved, and to bring to a satisfactory conclusion what has been begun.[3]

2

The Cambridge lecture and subsequent discussions were a resounding personal success for Kammerer, although those who held strong neo-Darwinian views remained of course sceptical. But even the latter came out of the meeting convinced of his sincerity. Thus the Hon. Mrs. Onslow, formerly Miss Muriel Wheldon, a lecturer in zoology and friend of Bateson's, gave the following account of the meeting, in a letter written to him the next day (Bateson himself refused to attend—see below).

Dear Mr. Bateson,

I went to Kammerer's lecture. There was a demonstration of his specimens in the afternoon. I did not get back to Cambridge in time for that though I saw a few at the evening lecture.

I did not in the least know what kind of man to expect. I had never been particularly interested in him or his experiments. I was most favourably impressed by his personality. I thought him delightful and he appeared quite sincere, genuine and very much in earnest. He read a straightforward account of his experiments in quite good English. I thought it a praiseworthy effort to prove his hypothesis. It dealt only with Salamandra, Toads, Proteus and a creature with siphons. But I felt very strongly indeed that, given *all* the data and the necessary knowledge, I should, if I performed the experiments myself, put a quite different interpretation to it.

The lecture was given in the anatomy theatre. It was well attended by young men and women though not full. There were present also the Professor of Zoology,* MacBride, Scott, Keilin, Balfour-Browne, Lamb, Potts, Gray, Haldane, Gadow, Hopkins and Miss Saunders. That's about all.†

After Kammerer had finished, Gadow got up and criticised the experiments. He cast doubts on most of them. MacBride gave them all the highest praise: in fact he was Master of the Ceremony and had been all day, I gather. All that was said about Mendelism was quite favourable,

---

* Professor J. Stanley Gardiner, an enthusiastic supporter of Kammerer.
† Actually, there were many more prominent scientists present, as we shall see.

or perhaps fair, and in order, except from MacBride who said that it had not thrown the slightest light on the differentiation of species nor on evolution in general. This was loudly applauded, almost solely by your son; so that you can deal with that.

She goes on for a while about the stupidity of students, then turns against their teachers:

The Professor of Z. very strongly urged the youth of this generation to take up K's work without delay and either to prove or disprove it.

Kammerer himself made a most valiant effort to follow his critics, and he succeeded with conspicuous skill and dignity, though his English was not good enough to enable him to reply. He therefore spoke in excellent German, I mean clear enough for me to follow to some extent. Gadow interrupted at intervals, the situation being extremely comic.

The speeches made by Gadow, the Professor and MacBride afforded, only too painfully, evidence for the very low grade of scientific effort exhibited by the schools of Zoology in the two greatest Universities of this country—that of Gadow being particularly coarse and of the lowest order possible.[4]

The Hon. Mrs. Onslow certainly had a sharp tongue, yet Kammerer came off best of the lot. Bateson replied curtly:

Dear Mrs. Onslow,
Thanks for interesting account of Kammerer's meeting. I was well out of it. That the performances of the Zoological leaders were deplorable, I can readily believe.

Since the Cambridge visit provides, as we shall see, the most important clue to the whole riddle, I have collected first-hand reports from the surviving eyewitnesses. Nearly fifty years after the event, they disagreed of course in some detail, but on the general impression Kammerer made their recollections are both vivid and unanimous.

Thus Professor W. H. Thorpe writes (May 25, 1970): 'I was at the lecture and it has remained one of the strongest impressions of my undergraduate career.'

G. Evelyn Hutchinson—now Sterling Professor of Zoology at

Yale—remembers some relevant details (letter dated April 17, 1970):

The lecture was given in German,* very Viennese, and was the first time that I realised how beautiful the language could be. It was interpreted by old Hans Gadow whom we all loved [Dr. Gadow was an eminent reptile expert]. A day or two later the annual conversazione of the Society took place. Kammerer exhibited his *Alytes* with one horny pad in place. The other was allegedly sectioned; the sections being also on view. Kammerer claimed that these sections, which were I think indeed made from an amphibian nuptial pad, differed in detail from those of any other known species. Of course no one in Cambridge was an expert on this matter. The toad specimen was taken out of the museum jar and examined under a binocular microscope. Unfortunately I, as one of the Council members, had certain duties, I forget what, and although I was on the edge of the crowd round the binocular, never got to look down it. J. B. S. Haldane, however, did, and I remember that he said he could see the characteristic ridges of a nuptial pad. I had a long talk with J.B.S., in Cornell, just before he died and asked him if he could remember the incident, but unhappily he could not. If the ridges were there, the specimen was not the one that was later found to be faked. I still think that it would be worth while for someone with a passion for breeding toads to repeat the *Alytes* experiments.

Kammerer stayed with Dr. E. J. Bles who was a great friend of mine. Bles had been the first person to breed the African clawed toad *Xenopus* in captivity. Bles had tried to breed the cavernicolous *Proteus* unsuccessfully and Kammerer had succeeded. He told me after Kammerer had left that he was convinced that he was absolutely honest. Bles was a philosophically neutral Darwinian who would not have been partisan about uncertain issues.

Professor Quastel (see above, pp. 71–2) writes as follows:†

I am just wondering how to reply to your letter about Paul

* It was actually given in English; Kammerer spoke German only in the discussion.
† August 26, 1970.

Kammerer. The incidents happened about forty-seven years ago and all I can remember distinctly now are the impressions I had of Kammerer as a person and as a scientific man. When I visited Vienna in 1923 I was a very young man, newly embarked in the field of biochemistry, and certainly not a trained biologist, though I knew enough to realise the possible importance of Kammerer's work. I was a member of the Cambridge Natural History Society and I took a neutral stand between the two opposing groups concerning the inheritance of acquired characters, one headed by MacBride, under whom I had studied a little zoology at the Imperial College of Science, and the other headed by Bateson. It was thought that if Kammerer would bring his critical specimen of *Alytes* to Cambridge and give a talk about it to all the English scientists interested, some progress might be made towards resolving the differences between the two groups of biologists.

I was asked, therefore, to visit Kammerer in Vienna, to bring back relevant photographs, but above all to see if Kammerer would visit Cambridge, give us a lecture and show us the critical specimen.

I remember meeting Kammerer in Vienna very well and feeling that he was a man of great charm and integrity. I remember receiving the photographs of *Alytes* from him, possibly also the negative, but I am afraid I do not recall being present in the studio where the photographs were made. If Dr. Przibram states I was there, I suppose I must have been there, but I do not recall this now. I do not think that I worried very much about this matter, however, because Kammerer gave me his assurance that he would visit Cambridge and bring his critical specimen with him.

I remember giving the photographs (and possibly the negative) to my good friend Michael Perkins when I returned to Cambridge and Michael played an important part in arranging for Kammerer's lecture and demonstration. (As you probably know, Michael Perkins died a few years later, during a 'flu epidemic in London.) I met Kammerer when he landed in England, and translated his lecture, which was in German, into English. The critical specimen of *Alytes* was, so far as I can remember, in Kammerer's hands until the day of the demonstration.

A large crowd attended Kammerer's lecture and demonstration. However, Bateson and a few other important biologists who we hoped would be present, did not attend. Many people examined the critical specimen closely. I never heard any statement, from those who saw the specimen, that it was faked. I feel sure that there were sufficiently reliable scientists present to give an unprejudiced opinion. The general feeling among my friends was that the demonstration was a success and I, personally, at that time, could not have believed that we were all being hoodwinked. Moreover, Kammerer's personality was such that those of us who knew him, even for only a short time, could not believe that he would deliberately deceive us. I met him again in Austria, after the Cambridge visit, and my faith in him as an honourable man never wavered. I was willing to believe that he—like any of us—might misinterpret his results, but I could not believe that he himself would fake them.

My faith, however, was later shaken by his suicide. Would a man who had given so much of his life to this work and who really felt his results were genuine, commit suicide knowing what interpretation would be given at this critical juncture, by the scientific world, to his suicide? I know that this line of reasoning may not be fair, because Kammerer's suicide may have been due to reasons other than scientific—but, whatever the reasons, there is no doubt that all who questioned Kammerer's work, now felt themselves justified. And I feel that I must agree also with this verdict.

I cannot believe that Kammerer himself made the injection but I have no idea who could have made it. I am undecided as to whether the injection was made before the Cambridge visit—if it was, it must have been very cleverly done to deceive all the biologists who saw the specimen in Cambridge. I personally was no expert and could not give an opinion, but my biologist friends all assured me that the *Alytes* demonstration was genuine. It is conceivable that if it was doctored before the Cambridge meeting, it was further doctored at a later date—but I truly cannot bring myself to believe that Kammerer would do this.

The last but one paragraph is, of course, inferential, and seems to contradict the rest of the letter. But that is how most people felt about Kammerer's suicide. That—and the riddle of the 'injection'—I shall discuss later; at present we are only concerned with the eyewitness accounts of 1923, undistorted by hindsight. To complete the record, I shall quote Dr. L. Harrison Matthews, F.R.S., the former Scientific Director of the Zoological Society:*

I am interested to hear that you are writing a study of Paul Kammerer, and I well remember his visit to Cambridge.

I cannot quite remember why we invited him, but I think it was because there was considerable controversy about his reported results, and we, as is the habit of the young—an excellent one too—felt that he was not getting a fair deal from his opponents, and that it would help him to come in person and say his say. There were some post-war money difficulties and we raised funds for entertaining him and, I think, to pay his fare.

Anyway, we [Michael Perkins, J. H. Quastel and L. Harrison Matthews] went to Harwich in the evening and met K. somewhere short of midnight. (Though it was late in April, it was a raw, misty night and we all wore overcoats or mackintoshes). We gave Kammerer a snack at the Parkeston Quay Hotel, and travelled on some night train to Cambridge. On the journey we naturally kept him busy with questions and had a pre-view of his specimens. I have forgotten where we parked him . . . we probably got him a room in one of the colleges (perhaps Trinity, Perkins'). [Before the formal lecture] a demonstration of his specimens was put on in the old zoology lab. on the top floor of the old building. He had his *Alytes* and micro-sections of the alleged pads on the hands, the *Ciona* with elongated siphons, and the *Proteus* with eyes. He also showed photographs of the first and second generation *Ciona* with short and long siphons living in an aquarium, and he had the specimens of his salamanders. . . .

Kammerer struck me as a frank, open-hearted man, intensely interested in his experiments, and ready to have them subjected to any scrutiny or test. His personal appearance was in his favour, for he was a handsome chap, with a most friendly manner. Nevertheless we (I) evidently was not fully convinced, for I see that I labelled the photo-

* Letter dated July 24, 1970.

Professor Hans Przibram in 1924, age 50.

Paul Kammerer and Michael Perkins, Cambridge, 1923.

Kammerer's specimen of *Proteus* with eyes restored, exhibited in Cambridge and London, 1923. *Photo : B. Stewart, Cambridge.*

Below left : Top left: normal adolescent specimen at start of experiment. Top right: same specimen reared on *black* soil, fully grown, showing colour adaptation. Bottom left: offspring of above reared on *yellow* soil. Bottom right: offspring of above reared, like parents, on *black* soil. *Photo-montage : B. Stewart, Cambridge.*

Below right : Top left: normal adolescent specimen at start of experiment. Top right: same specimen reared on *yellow* soil, fully grown, showing colour adaptation. Middle left: offspring of above reared on *black* soil. Middle right: offspring of above reared, like parents, on *yellow* soil. Bottom centre: third generation animal reared, like parents and grandparents, on *yellow* soil, just after metamorphosis. *Photo-montage : B. Stewart, Cambridge.*

Sea-squirt (*Ciona intestinalis*). Normal specimen showing the two short siphons on top. Photographed in the sea aquarium of Prof. Cerny, Vienna.

Above: A group of sea-squirts in the process of regenerating their previously amputated siphons. A few of the regenerating siphons show already pronounced elongation and are more slender than the old stems. (Aquarium Prof. Cerny.)

Below: Non-amputated descendants of *Ciona* which, through repeated amputations, became transformed into the long-siphoned variety (*var. macrosiphonica*). (Aquarium Prof. Cerny.)

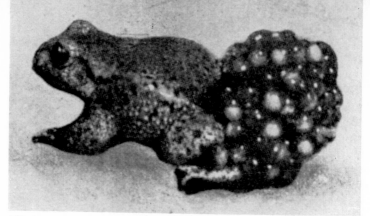

The midwife toad *Alytes obstreticans*, male, with egg-strings tied around hind legs. (*Neuvererbung*, p. 19.)

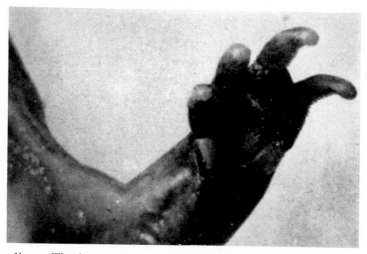

Above : The last specimen of the fifth generation of 'water-bred' *Alytes*, photographed in the Reiffenstein studio, Vienna, 1922. The horny spikes on the outer edge of the hand are clearly visible. (*Neuvererbung,* p. 20.)

Below : Top: microtome section through nuptial pad of fifth generation 'water-bred' *Alytes*.

Below : Bottom: microtome section through the same area of normal (land-breeding) *Alytes* during the mating season.

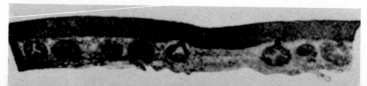

graphs when mounting them (either in 1923 or 1924) 'to show alleged inheritance of acquired characters' and 'showing inheritance (?) of acquired characters'.

I have always found it difficult to believe that K. was a charlatan; he may have deceived himself through being insufficiently critical, or he may have been deceived by his assistant, as is said. The whole controversy was full of muddled thinking on both sides, and old MacBride's ridiculous ex-cathedra statements did not help. Looking back, it is surprising that a little clear thinking did not ask half a dozen key questions which would have sent K. back to his lab. to find the answers which would have been final. Whether his suicide was due to being found out, or to his finding out that he had been tricked, you probably know better than I. From what I saw of Kammerer, I liked him. I should like to say again that I think that Kammerer was honest though his results were 'phoney'. If there was deception he was one of the victims. From what I saw of him I feel sure that he was sincere and really believed what he said.

Dr. Matthews then went on to pay a moving tribute to the abilities of Michael Perkins, whose name we shall encounter repeatedly (he died of septicaemia at the age of thirty-three).

Neither Gadow nor Boulenger or any other sceptic present at the Cambridge meetings and demonstrations expressed doubts regarding the authenticity of the specimens or the microscopic sections. This is the crucial point to be retained.

3

In spite of his worries, Kammerer seems to have enjoyed himself in England. There were various anecdotes going around concerning the distinguished visitor. The Hon. Ivor Montagu, who studied zoology before he turned to politics and films, writes in his autobiography:

When Kammerer came . . . he was only the second scientist of 'ex-enemy' nationality—Einstein being the first—to visit Britain after the 1914-18 War. I asked him down to Townhill [the Montagu's country place] and pressed him to say if there were anything he specially wished to do or see before he left the country. He had two wishes: to eat a kipper and to meet Bernard Shaw. He explained the first by saying that an old English lady, whom the war years had trapped in Vienna, had above all missed during her exile the taste of kipper and made him promise to eat one for her. The second was as easy to satisfy for by then I had met G.B.S. and knew he had absurd ideas on evolution.[5]

Gregory Bateson has another anecdote:

I was part of the little undergraduate biological club which got Kammerer to Cambridge in 1923. My most vivid recollection is of his telling the story of how on the previous day he had stopped at the London zoo and stood in front of the llama. He was looking at the llama and thinking that the llama's face resembled that of William Bateson. At that moment the llama spat.[6]

But the sequel was not quite so funny.

4

One conspicuous feature of the Cambridge meeting was the absence of William Bateson. He had clamoured for a specimen of *Alytes* to be sent to him for examination in Cambridge; now that it had been publicly announced that a specimen was to be exhibited in Cambridge, Bateson refused to examine it. The reason he gave for his refusal to attend (in a letter to *Nature*, June 2, 1923) is somewhat odd. After reiterating yet once more that he had challenged Kammerer in 1910—thirteen years earlier—to produce a specimen of *Alytes*, and that Kammerer at the time had failed to do so, Bateson continued:

But one specimen (presumably that photographed) was known to be preserved in Vienna. It had been examined by visitors to the *Versuchsanstalt*, who reported verbally and variously as to what they had seen. A few weeks ago the announcement was made that this *Alytes* would be shown in Cambridge, and I received an invitation to attend a meeting at which it would be exhibited. Knowing that Dr. Kammerer had abstained from appearing at the Congress of Geneticists which met at Vienna in September last, I inferred that he had no new evidence to produce, and I therefore excused myself from attendance, not wishing to enter deliberately into what was likely to prove a profitless altercation.

However, the Cambridge meeting created such a stir in the Press and among scientists that the Linnean Society of London invited Kammerer to repeat the lecture on May 10. This time Bateson, who was a prominent member of the Linnean, could not refuse.

The annual Proceedings of the Society gave a short and fair summary of the lecture—which had already been published in full in *Nature*—and devoted four paragraphs to the discussion. The first reads:

Dr. Bateson, f.r.s., having complimented the lecturer on his enthusiastic devotion to his subject, dissented from several of his conclusions.

That is all about Bateson. The other two disputants mentioned in the Proceedings got twelve and seven lines respectively. They were Mr. J. T. Cunningham and Professor E. S. Goodrich, f.r.s., Secretary to the Linnean Society. They did not question Kammerer's results, but objected to some of his interpretations. The fourth and last paragraph reads:

The lecturer replied, Prof. MacBride acting as interpreter. He submitted that the criticisms of Dr. Bateson and Mr. Cunningham were irrelevant, and remarked that control experiments would be carried out in Cambridge. The number of individuals subjected to his experiments varied from as few as twenty to as many as a hundred in different cases.

Bateson refrained from any open expression of hostility. According to MacBride, writing in *Nature* a fortnight later,[7] 'Dr. Bateson completely withdrew his charges of bad faith on the part of Dr. Kammerer, and accepted his published results as genuine, claiming, however—as he had the full right to do—to differ from the deductions which Dr. Kammerer drew from them.' Kammerer also confirmed that 'he [Bateson] expressly apologised to me in case I had considered his previous attacks too rude'.[8]

It stands to reason that MacBride and Kammerer could not publish versions of what had happened at the meeting which did not correspond to the facts—particularly in *Nature*, which all participants at the meeting were bound to read. Nor did Bateson dispute his apologies and retractions. But no sooner had

Kammerer left England than Bateson resumed his hardly veiled accusations of fraud.

Kammerer left on May 11. On May 16 Bateson wrote a long letter to *Nature*, comparing one of the photographs Kammerer had shown to 'spirit photographs like those handed about a few years ago'. He also asked Joan Procter, in charge of reptiles at the British Museum (Natural History), for advice how to feed *Alytes*, and tried to recruit her for the anti-Kammerer campaign. Her answer was discouraging: 'I only wish that we had a richer material to draw upon, and that I could have helped in any way to clear up the *Alytes* question. . . . I haven't the least objection to being drawn into war with Kammerer but, as I said before, have not been able to be of the slightest real assistance.'[9]

Though it sounds difficult to believe, Bateson did not avail himself of the unique opportunity at the Linnean meeting to make a thorough examination of the specimen which he had been so anxious to see. He did not even ask for it to be taken out of its museum jar so as to examine *both* sides of the hands, as his colleagues in Cambridge had done—for in the jar only one side, the palmar side of the hands, could be seen. When Kammerer subsequently challenged him on this point,[10] Bateson replied (September 15):

> The question remains, what is the real nature of the swellings in the animal exhibited? That on the palm did not look like a nuptial pad. What there may have been on the back of the hand I do not know. I made no statement about it, though Dr. Kammerer says I did. I might no doubt have asked to see the back, but I had no reason to suppose there was anything more to see.[11]

It was the same reasoning which made him decline to go to Cambridge because he 'inferred' that Kammerer 'had no new evidence to produce'. Yet we remember his instructions to Mr. Martin (p. 70): 'Note very carefully the position of the alleged *Brunftschwielen* on EACH side. Make them take it out of the bottle. Ask to have the specimen out in a dish where you can *examine it quietly with a dissecting lens at your leisure.*'

At the Linnean meeting Bateson was the most important person present. He had a dissecting lens and all the leisure to look at the specimen from both sides. Why did he let this unique opportunity pass? Could it be that he looked away for fear of being convinced? Kammerer implied that much:

> [If Bateson had been willing] it might have been possible for me to make him see what he did not wish to see; I would certainly, for his benefit, have removed the *Alytes* specimen from the jar, and he would have been able to view it—without obscuration by glass or background—from all sides under the lens: I treated it this way during my stay in England for many colleagues (as, for example, for Mr. E. G. Boulenger and Sir Sidney Harmer).[12]

Bateson's letter of September 15, in which he explained that he did not ask to see the back of the hands because he had 'no reason to suppose there was anything more to see', ended with a dramatic surprise:

> I have a strong curiosity to see this *Alytes* again. For the opportunity of examining it at leisure in the British Museum where comparative series are available or if preferred in Professor MacBride's laboratory, I am willing to pay twenty-five pounds either to the *Versuchsanstalt* or to other appropriate authority. Plenty of responsible people travel between Vienna and London, and there should be no difficulty in arranging for safe conveyance.

Thus, having declined to examine the specimen properly when it was in London, he now developed a 'strong curiosity' to see it again, and asked that it should be sent to London once more. Being aware of Kammerer's reluctance to expose his last, battered specimen to further damage, he must have counted on his request being refused. So it was; but Kammerer's reply[13] was, under the circumstances, remarkably restrained:

> The type specimens of my experiments are in the Museum of Experimental Development attached to the Biological Institute of Vienna, and are the property of the Museum. I communicated Dr. Bateson's proposal to the directorate, and added, as my own opinion, that I was not in favour of exposing the critical specimen of *Alytes* with nuptial pads to the dangers of a second journey, only because Dr. Bateson had neglected the opportunity of examining it when he was able to do so. Nevertheless I did not oppose a veto to the directorate sending the specimen, if they wished to do so.

Przibram, in his capacity as Director of the Institute, wrote shortly afterwards:

Vienna XIII/7 Hiezinger Hauptstrasse 122.
My dear Professor Bateson,
Having read your offer about Kammerer's *Alytes* in *Nature* No. 2811, my proposal is this: that you may carry out your previous intention of coming to Vienna yourself. I would gladly renew my invitation to you to spend some time at my house. Thus you would be given ample opportunity to examine the specimen without risk of its loss. It was mainly my wish to satisfy you that made me consent to Kammerer taking the specimen to England. I am sorry you have not availed yourself of this opportunity, but I could scarcely take the responsibility of entrusting the unique sample to anybody else (I had in fact declined to do so on a previous occasion, as M. Boulenger will affirm). It would indeed be a great pleasure to see you with us. Believe me, dear Professor Bateson, most sincerely, your old friend,

Hans Przibram

Bateson published Przibram's letter in *Nature* (December 22), together with his own reply:

I was not without misgiving that difficulties might be raised. For that reason I offered a sum, twenty-five pounds, calculated to cover the railway fare, ten pounds, of a special messenger, with a sufficient margin. I understand that the obstacle is not financial or I would gladly now double my offer.

I might no doubt have been a little quicker, but in amends, and in the hope of bringing the matter to a definite issue, I made the offer, not an unfair one, which you have declined. Yours truly, W. Bateson.

And there the controversy was to rest for three years until the final tragedy in 1926. But during those last three years Paul Kammerer was a broken man.

Eight

I

I have mentioned the financial ruin of the Kammerers and Przibrams. The term 'inflation' as applied to the relatively moderate increases in the cost of living after the Second World War has no relation to the catastrophic events in Central Europe in the early 1920s. The Austrian Krone was originally worth the equivalent of a Swiss franc. By the end of 1920 it was worth one centime—one-hundredth of a franc. By the end of 1921, 1,400 Kronen bought a franc; in 1922 the figure had again increased tenfold. Within a few months the Kammerers' family fortune had melted away. Worst hit, of course, were the white-collar workers, including academics, reduced to starvation salaries. Kammerer's amounted, in 1923, to the equivalent of £150 a year. Respectable middle-class housewives prostituted themselves; elderly magistrates could be seen queuing up at soup kitchens run by charities. Hans Przibram, as we remember, had founded and financed the Biological Institute largely out of his own means, until the Austrian Academy of Science took it over in 1914. After the war he wrote pathetic letters to his one-time friend Bateson, offering rare books from his private library for sale, in exchange for British scientific publications, which the Institute could no longer afford to buy.

MacBride, the pugnacious Irishman with a heart of gold, after the Linnean incident appealed privately to Bateson— disregarding the public controversy in which they were involved —trying to persuade him to show a more human attitude to Kammerer:

I want merely to put before you the economic, material difficulties which exist and which would not occur to an Englishman.

Kammerer is going to America in the autumn. He will doubtless bring his critical specimens with him. I know that he has the utmost difficulty in raising money—in particular I have been told of his gratification at receiving a grant of £50 which would keep him till the autumn! I have offered to put him up, if he passes through London, but I fear that this should prove too expensive in addition to his trip. If he should pass through London, the opportunity which you ask for the leisurely examination of his *Alytes* specimen would be afforded.

May I ask, have you taken steps to send Kammerer a copy of your last letter [to *Nature*]? If you have not, you must not interpret silence as an inability to answer. *Nature* is an expensive luxury in Austria—it costs about 18,000 Kronen a copy and is far beyond the purse of the average scientist. Kammerer would have seen nothing of the criticisms of his lecture published in *Nature* if his friends had not sent him copies of that journal.[1]

By that time Kammerer was already in the United States on a lecture tour. He had resigned from his post at the Biological Institute to support his family by journalism and lecturing. The £50 that MacBride mentions represented the golden handshake of the Austrian Academy of Sciences after nineteen years of services rendered to the *Versuchsanstalt*.

2

The popular lectures in the United States and on the Continent were immensely successful, but definitely harmful to his scientific reputation. He captivated the public by his charm and obviously sincere belief in his theories—regardless whether his audience consisted of Cambridge biologists or New Yorkers without scientific training. But the promoters and the Press created around the lectures an atmosphere of sensationalism which he was partly unable, partly too naive, to stop.

The ballyhoo had started already at Cambridge. On April 24, a week before Kammerer's lecture, the Council of the Natural History Society convened 'to consider the question of notifying the Press of Dr. Kammerer's lecture'.[2] According to the minutes of the meeting, 'a heated argument ensued' on the question whether or not to invite the Press. The Hon. Ivor Montagu, a member of the Sub-Committee in charge of arrangements for Kammerer's visit, argued that 'although it was understood that Dr. Kammerer did not wish that publicity should be given to his intended visit, circumstances had arisen which the Sub-Committee had considered to justify the appearance of reports of the lecture in the Press'. What these circumstances were, the minutes do not say. In the end a motion, proposed by Evelyn Hutchinson, to the effect that 'the Council will not allow any Press report of Dr. Kammerer's lecture without their consent', was put to the vote and carried unanimously. After that it was decided that an 'official report of the meeting' was to be sent to *Nature*, 'it being clearly understood that no official judgment should be passed on the merits of Dr. Kammerer's work'.[3] The upshot of it was: in accordance with Kammerer's wishes, no Press publicity, except for a report to *Nature*.

However, during that Council meeting 'a further complication arose in that a member of the Sub-Committee had invited as a private guest a person who as well as being interested in the inheritance of acquired characters was a member of the staff of a daily paper. This person had given proof of an intention to send a report of the meeting to his paper. The President explained that the meeting was a private one of a private Society and that should it be the wish of the Council that no unauthorised report should appear in the Press, then it would be the duty of the member in question to ask his guest not to violate the hospitality shown to him by publishing a report of that Meeting.'[4]

The member of the Sub-Committee in question was Michael Perkins, the Curator of the Society; and the sinister 'person' whom he had rashly invited was a reporter on the *Daily Express*.

The sequel was catastrophic. The reporter came to Cambridge, but was debarred from Kammerer's lecture, to which only members of the Society and invited guests were admitted. This secretiveness whetted the reporter's appetite even more, and convinced him that sensational news was being hatched behind the closed doors. He managed to talk to members of the Society and perhaps to Kammerer himself, and to collect a few hints about the restoration of sight to the blind newt *Proteus*, the changed breeding patterns of amphibians, and so on. The result was a sensational front-page story in the *Daily Express* on May 1. The headline WONDERFUL SCIENTIFIC DISCOVERY was splashed over all six columns in the early-morning edition. Then came the sub-titles:

EYES GROWN IN SIGHTLESS ANIMALS
Scientist Claims To Have Found How
To Transmit Good Qualities
Hereditary Genius
Transformation of the Human Race

But the sub-editor of the late morning editions obviously thought that these headlines were too restrained, for in the late London edition, though the text remained unaltered, the headlines were thus improved:

RACE OF SUPERMEN
Scientist's Great Discovery
Which May Change Us All
Hereditary Genius
Eyes Grow In Sightless Animals

The minutes of the next Council meeting of the Natural History Society read:

The minutes of the previous meeting were read, and, with certain corrections necessitated by the somewhat hurried and undignified departure of the proposer of a motion at that meeting, were confirmed. As the result of a series of most unfortunate circumstances, culminating in the appearance of an inaccurate and unauthorised account in the *Daily Express* of Dr. Kammerer's lecture at the Society's meeting of April 30th, Mr. Perkins tendered his resignation from the Council and retired from the meeting. The consideration of the acceptance of his resignation was deferred until the other business was completed.

The Council considered that the best course to take in respect to the inaccurate and unauthorised account of Dr. Kammerer's lecture in the *Daily Express*, was to ignore the *Daily Express* entirely. In view of the fact that Mr. Perkins had done his best to prevent an unofficial report appearing in the *Daily Express*, the President proposed that 'The Council do not accept Mr. Perkins' resignation'. Seconded by Professor Gardiner. The motion was carried unanimously.[5]

But the damage was done. The *Daily Express* story was cabled to the United States before Kammerer's arrival. The results were again headlines, such as those which appeared in the *New York World*, May 5, 1923:

VIENNA BIOLOGIST HAILED AS GREATEST OF
THE CENTURY
Proves a Darwin Belief
Theory Wins Recognition From Cambridge
University Scientists

The headlines were actually not as dotty as they may seem; for the cabled despatch quoted Professors Nuttall and Gardiner; and G. H. F. Nuttall, Professor of Biology at Cambridge, did actually say that Kammerer had made 'perhaps the greatest biological discovery of the century'; and Stanley Gardiner, Professor of Zoology, did say that 'Kammerer begins where Darwin left off'.

Even the staid *New York Times* was carried away:

SCIENTIST TELLS OF SUCCESS WHERE DARWIN
MET FAILURE
Eyes Developed In Newts
Demonstrates Acquired Qualities May Be Inherited
Austrian Savant's Laurels
Evolution Would Be Speeded Up If Best
Characteristics Could Be Transmitted

There was also a portrait sketch of Kammerer with the caption 'Hailed as a second Darwin'. The text of the article, however, spread over five columns, gave a fairly accurate summary of the experiments.

3

This avalanche of publicity preceded Kammerer's arrival in the States. The lecturing agency which organised his tour, in its announcements, exploited it to the hilt—particularly the statements by Nuttall and Gardiner. The American academic world, generally not averse to a little publicity, professed to be shocked.

The most indignant voice was, predictably, that of the great T. H. Morgan, Professor at Columbia University, pioneer of the chromosomal theory of heredity, whose authority in American genetics was similar to Bateson's in England. When Morgan was invited to join the committee sponsoring Kammerer's visit he wrote back:

Dear Sir,

I should under no conditions consent to become a member of the committee about which you write, if the kind of announcement that you sent me is intended as a statement of Dr. Kammerer's qualifications and accomplishment. I have, of course, no personal objections to Professor Kammerer, and have been long familiar with his work— both as to results and deficiencies, but this kind of advertising is the sort of thing for which no man interested in real scientific development in America should for a moment stand sponsor.

Very truly yours

T. H. Morgan[6]

He sent a copy of this letter to Bateson. Several months later, on the occasion of Kammerer's second American lecture tour, Morgan again wrote to Bateson:

Dear Bateson,

I am wasting your time by sending you a little more of the Kammerer developments. He is running true to form— serving one good purpose, at least: that of an indicator. He divides our contemporaries into two rather well defined groups; this will help ultimately to clarify the situation.[7]

The two well-defined groups were, of course, the neo-Darwinians and the neo-Lamarckians. But the division, far from 'helping to clarify the situation', led only to a hardening of attitudes and whipping up of emotions.

Yet Kammerer also found powerful allies in America. One of them was J. B. Watson of Johns Hopkins University, the founder of the Behaviourist school, which for the next fifty years was to dominate academic psychology. Watson wrote:

Professor Kammerer's work on the inheritance of acquired characteristics has startled the world. Biological students for more than a generation have accepted the view that acquired characteristics are *not* inherited. Professor Kammerer's experiments seem to point otherwise. His results are in the forefront of discussion today in biological circles. We all want to believe his facts if they are true. It means so much to the educator, to society in general, if they are true. American students are extremely fortunate in being able to hear this distinguished biologist at first hand, and to ask him questions about his technique and control.[8]

The success of the American lecture tour in the autumn of 1923 is indicated by the fact that Kammerer was invited to make a second tour in February, 1924. He was cold-shouldered by one of the two 'well-defined groups', but at least nobody accused him of faking his results. During his second visit, his book *The Inheritance of Acquired Characteristics* was published by Boni & Liveright in New York. One of the advertisements—lovingly preserved among Bateson's papers—read:

This is the epoch-making work which won for Dr. Kammerer the acclaim as a second Darwin. This work covers the successful experiments which solved the problem Charles Darwin despaired of, and which was universally held to be insoluble.

As professional writers know, publishers are not in the habit of consulting their authors about their advertisements. Nevertheless, all this did not help to endear Kammerer to the scientific community.

I have described in some detail the events at the Cambridge Natural History Society's Council meetings which led to the 'inaccurate and unauthorised' report in the *Daily Express*, and started the avalanche of publicity in the United States. My purpose was to show that—as the Council's minutes expressly state —'Kammerer did not wish that publicity should be given to his intended visit', and that he had no part in the sensational developments which followed. Other scientists, from Einstein downward, had become victims of sensationalism, without being branded as charlatans.

4

I have mentioned three factors which contributed to Kammerer's growing despair: financial penury; the relentless harass-
ment and veiled accusations of fraud by Bateson and his allies;
and the wrong kind of notoriety which indirectly played into the
hands of his detractors. An added problem were women.

All who knew him agree that he had an extraordinary attraction for the fair sex—as it was then called. Even the discreet
obituaries mentioned it. His daughter Lacerta returns to the
subject several times in her letters:

> He was so used to the magnetic effect he had on women
> that he did not try to make conquests—with very few ex-
> ceptions; he conquered too many for his own comfort as
> it was.

In 1920 the Kammerers spent the summer in an Austrian
lake resort:

> Two other women who had never met us before were also
> staying there. One was in her twenties and the other about
> forty. Both went obviously quite gaga about P.K. 'Forty'
> danced in the moonlight under Father's window! She even
> picked up a snake (a harmless smooth adder) for him. I can
> still remember the greenish hue on her face as she did it.
>
> I think my mother kept young women out of my father's
> way as much as it was in her power. And did he attract
> them! I remember streams of them trying to get 'in his
> presence' with all sorts of excuses.

And yet it would be quite wrong to think of him as a Viennese
Casanova. To call him a romantic Werther would be closer to
the mark. He did have affairs, but they were charged with an
intensity on his side which led to self-torture—Werther's *Leiden*.
There was that episode with Alma Mahler. Soon after her he
was to fall in love with a well-known, eccentric paintress called
Anna Walt. Felicitas was understanding: she agreed to a
divorce on the pretext of mutual incompatibility. Kammerer
married Anna Walt—and with her he had to endure a real incompatibility of temperaments, for the marriage lasted only a
few months. After one of their deadly rows he swallowed an
overdose of sleeping pills, but vomited them out. He went

94

through a period of depression, returned to live with Felicitas
and Lacerta and the Wiedersperg clan, who apparently restored
him to normality.* Felicitas took it very well: 'After Father's
return home at the end of the Anna Walt interlude I heard
mother say: "People admire me for being so forgiving, but if
you love a person you just love him and that's that." '

But a few years after this episode—about the time of his
return from America in 1924—he fell in love with another *femme
fatale*—about which later.

5

Penury, calumny, the wrong type of fame, his cherished collection of specimens—the material evidence of his life work—destroyed; and an unhappy love affair. . . . The sparse accounts of
his state of mind in those last two years of his life speak either
of a man utterly dejected or of one full of energy and gaiety.
Evidently the two states alternated in a manic-depressive cycle.
But even during the depressive periods he worked on doggedly.
Apart from popular articles and lectures, he worked on a book,
which even his detractors admitted to represent a classic con-
tribution to evolutionary biology. It was finished in December,
1925, and published in the summer of 1926, a few months before his suicide, under the unpromising title: *Der Artenwandel
auf Inseln und seine Ursachen ermittelt durch Vergleich und Versuch
an den Eidechsen der Dalmatinischen Eilande.*† It is dedicated
'To my daughter Lacerta for her eighteenth birthday'.

The book is based on three expeditions, 1909, 1911 and 1914,
the first on fishing trawlers, the second and third on the explorer
ship *Adria*, sponsored by the Austrian Academy of Science. He
visited about fifty, mostly uninhabited, islands, ranging in size
from fairly large ones to solitary rocks. These island abound in
lizards, and Kammerer was able to catch on each island twenty
to fifty live specimens and bring them home to the Institute,
where he used them for breeding experiments. His description
of the various ways of catching lizards is a delight. The methods

* By some legal quirk, produced by changes in Austrian law, the
marriage to Anna Walt could be annulled and the (Catholic) marriage
to Felicitas revalidated.
† *The Transformation of Species on Islands and its Causes, Ascertained
through Comparative and Experimental Work with Lizards of the
Dalmatian Isles.*

varied according to weather and the terrain; they included turning up loose stones; chasing the animals on open ground; using a modified butterfly net; using a fly on a fishing rod as bait, combined with a wire loop or net. Island lizards apparently vary in temperament from 'very shy' through 'trusting' to 'cheeky'. (One may also note, as incidental information, that the island lizards have established a sort of symbiosis with seagulls—the lizards are regular visitors to seagulls' nests, which they clean of parasites with the evident approval of the gulls.)

The study of island populations of animals which evolve in isolation is of immense importance to evolution theory. The tortoises and birds of the Galapagos archipelago, which vary slightly but significantly from island to island, provided Darwin with an important clue for his theory. Bateson had looked for evidence of Lamarckism in isolated Asian lakes. Kammerer's purpose was 'to follow in the footsteps of Darwin and Wallace . . . and, by studying the variations of lizards on different islands, to glean some insight into the causes of these variations —and thereby of the origin of species'.[9]

The two main species of island lizards—*Lacerta serpa* and *Lacerta fiumana*—do indeed show in different islands very marked variations in colour, size and shape. So much so that on Mali Bariak, a rocky islet near Lissa, he discovered the previously unknown variety which was named after him—*Lacerta fiumana Kammereri*. Granted that isolation facilitates or speeds up the emergence of new varieties, the problem is once more whether these new varieties are due to random mutations and natural selection, or to the direct influence of the environment. He came to the conclusion—as one would expect—that not chance but the nature of the environment, its temperature, humidity, lighting, fauna, etc., were responsible for the changes which started as individual adaptations and ended up by becoming hereditary. To prove his point he experimentally induced colour changes in his specimens—black to green or green to black—by altering their environment, much on the lines of his salamander experiments, and claimed to have shown that the changes did become hereditary.

But as for the evidence . . . Kammerer in 1923 handed over his collection to Dr. Wettstein, Vice-President of the Academy of Science, who wrote a classificatory survey of the material, which appeared as an appendix to Kammerer's book. Wettstein expressly notes that many specimens were in a very bad state of preservation as 'during and after the war they could not be given the necessary care'; all had lost their significant colours, and Wettstein had to base his classification on the coloured paintings which Professor Lorenz Muller, Curator of the Zoological Museum in Munich, had made from the live animals. The book was reviewed by MacBride in *Nature* after Kammerer's death.[10] MacBride remarks that the ruin of his collection was one of the causes 'which broke Kammerer's heart and drove him to suicide'.

Nine

I

The scandal which shook the scientific world and destroyed
Kammerer's reputation exploded in the columns of *Nature* on
August 7, 1926. On that day the journal published an article by
Dr. G. K. Noble, Curator of Reptiles at the American Museum
of Natural History, asserting that the famous specimen of *Alytes*
was faked.

Noble was thirty-two, a graduate of Harvard and Columbia,
who described himself in the American *Who's Who* as a
'Curator and Explorer'. He had led several naturalist expeditions to Guadaloupe, Newfoundland, etc. Gregory Bateson describes him as a ruffian, and among English zoologists who knew
him he had the same reputation. But he was without doubt an
expert on reptiles.

Noble had for some time been involved in the anti-Kammerer
campaign. The previous year, at the Annual Meeting of the
British Association, he had criticised Kammerer's experiments
on the grounds that the glands on the nuptial pads did not look
as he thought they should. His attack attracted little attention,
it was not reported in *Nature*, and neither Kammerer nor even
the faithful watchdog MacBride bothered to answer.* But early
in 1926 Noble visited the experimental Institute in Vienna, and
obtained permission from Przibram, with Kammerer's consent,
to subject the famous last specimen of *Alytes* to a thorough
examination. And this time he struck gold, so to speak. The

* MacBride did, however, answer Noble's criticisms referring to the
glands at a later stage of the controversy (*Nature*, August 21, 1926).

examination proved beyond doubt that the specimen showed no nuptial pads, and that the black colouration of its left hand was not due to natural causes, but to the injection of Indian ink.

Side by side with Noble's report, *Nature* published another one, by Dr. Przibram. He was a loyal friend of Kammerer's, but also a scientist of integrity. He repeated Noble's chemical tests, and confirmed Noble's findings.

Though the two reports agreed as to the evidence, they radically differed in their conclusions. Noble wrote: 'It has therefore been established beyond the shadow of a doubt that the only one of Kammerer's modified specimens of *Alytes* now in existence lacks all trace of nuptial pads. The question remains: might not this specimen at one time have possessed them?' And he concluded: 'Whether or not the specimen ever possessed them is a matter for conjecture.' Although he did not explicitly say so, he made it unmistakably clear that he did not believe this specimen, or any other specimen of *Alytes*, to have ever possessed the pads.

Przibram's report dealt first of all with the absence of the 'asperities', the rough skin equipped with spines, which is the main feature of nuptial pads, and offered a plausible explanation:

It is clear from the foregoing account [Noble's account] that the only one of Kammerer's experimentally modified *Alytes* still preserved cannot in its present state be regarded as a valid proof of the nuptial pads artificially produced in this species. We must endeavour to decide if the state the specimen is in now agrees with the state at the time of its preservation and before. The specimen is poorly fixed and preserved. Moreover, the epidermis is in several places ready to be shed or even shedding. It is a known fact, as Professor Franz Werner of Vienna asserts, that during repeated handling and shaking, the nuptial asperities get lost easily. The specimen has made the voyage to England and back again, and it does not look the better for it. Fortunately, there are photographic plates in existence showing the state of the specimen before it left Vienna for Cambridge, and during its stay in England.

After describing the photographic evidence he turns to the incriminating black fluid:

Whilst it is possible to come to a probable solution with respect to the spiculae [i.e. the absence of spines], we have not been able to elucidate the origin of the black substance. It is clear that it has nothing to do with the black pigment often seen in nuptial pads. The only possibility we can think of is that someone has tried to preserve the aspect of such black nuptial pads in fear of their vanishing by the destruction of the melanin [natural pigment] through exposure to the sun in the Museum case, by injecting the specimen with Indian ink. Kammerer himself was greatly astonished at the result of the chemical tests, and it ought to be stated that he had been asked and had given his consent to the chemical investigations.

Kammerer shot himself six weeks after the publication of the Noble report. He did not reply to Noble. He had a deep aversion for public polemics* and must have thought that Przibram's reply said all that there was to say; to protest his innocence would have been undignified. Only in his farewell letters did he broach the subject. One of them was addressed to Przibram, who conveyed its message to *Nature*:

> This sad end to a precious life may be a warning to those who have impugned the honour of a fellow worker on unproven grounds. It is in fulfilment of a wish expressed by Kammerer that I beg the Editor of *Nature* to publish his last word on the much debated but not solved question of a particular one of his specimens. Having convinced himself of the state it is in now, Kammerer alleges that someone must have manipulated it; he does not allude to a suspicion who this might have been.[1]

2

In view of the crucial importance attributed to Kammerer's experiments by friend and foe alike during his lifetime, it is strange that no biologist or historian of science has thought it worth while to study the evidence, instead of relying on hearsay and legend.

* He answered Bauer's, Bateson's and Boulenger's criticisms after a delay of six years. At the height of the controversy, Bateson alone had published five letters in *Nature*, not to mention other critics; Kammerer only two. He had left it to MacBride to answer the attacks.

The first question that arises is evidently: did Kammerer himself inject the Indian ink; if not, who else might have done it and to what purpose? The second question is whether the critical specimen did show the pads before it was tampered with. I shall start with the second.

Evidence for the reality of the pads, still in existence, is provided by the photographs of sections through the skin. We remember that in 1920 Kammerer sent photographs of these sections to Bateson (p. 70) and that Kammerer displayed the slides both in Cambridge and at the Linnean meeting. In Cambridge, B. Stewart made photographic enlargements of them, which Professor W. H. Thorpe has kindly made accessible to me; (see plates)*. Kammerer also sent slides, among others, to the American biologist, Dr. Uhlenhuth, who showed them to Noble; and finally Noble himself was able to inspect the original microscope preparations at the Vienna Institute. Noble had to admit that 'both sets of preparations agree fully with the descriptions given by Kammerer'.[2] But, he continued, these sections, which Kammerer claimed to have taken from the alleged pad of *Alytes*, looked rather similar to sections through pads from the frog *Bombinator*. He concluded: 'Kammerer's sections fall within the range of variability shown by the genus *Bombinator* (more properly called *Bombina*).'

He left it at that, but the inference was clear: Kammerer had made a section from the normal pad of an ordinary frog and pretended it came from *Alytes*. It was the same suggestion Bateson had made to Boulenger (see above, p. 71). But it was untenable on two grounds. In the first place it presupposed a conspiracy involving not only Dr. Przibram—whom Bateson, Boulenger and even Noble implicitly trusted—but also the histologist who actually took the microtome sections—Miss Olga Kermauner, sister of Professor Kermauner, University of Vienna.[3] A surreptitious injection of ink by a single individual is one thing; a conspiracy of at least three people, including the Head of the Institute, who send out forged slides to scientists all over the world, display them at lectures and exhibit them in their collection, is quite another.

The micro-photographs of the *Alytes* pads were, and are, available to experts for comparison with pad sections from other

* The same photographs appear in Kammerer's books *The Inheritance of Acquired Characteristics* (N.Y. 1924) and *Neuvererbung* (Stuttgart, 1925).

species of frogs and toads. The latter can be seen in the classic work of Lataste[4] and others. Nobody except Noble had suggested in public that the sections might have been taken from *Bombinator* ('fall within its range of variability'). No expert in the field backed him up. Bateson, dead by now, had voiced suspicions in private, but did not attempt to prove them. On the contrary, he came very near to retracting his veiled accusations in a letter to the younger Boulenger, written in August, 1920, shortly after Kammerer had sent him the microtome slides: 'Of course the sections cannot prove that *Alytes* treated in certain ways develop *Brunftschwielen*. Nevertheless they do constitute a puzzle. I doubt whether *Bombinator* or other species with the fully developed *Schwielen* can go through a stage like that seen in the sections.'[5]

Boulenger junior also refrained from commenting on Noble's report. He may have, earlier on, shared Bateson's suspicions about the origin of the slides, but after his visit to Vienna in the winter of 1922–3 he gave them up (cf. p. 71).

Apart from Kammerer's own, the only detailed histological description of the sections, and their comparison with sections from other species, was written by Michael Perkins in answer to Bateson's 1923 attack on Kammerer. An extract from Perkins' letter appears in Appendix IV. Perkins' conclusion was that the *Alytes* sections showed certain resemblances to two other species (*Bombinator* and *Discoglossus pictus*), but also characteristic differences from them. Bateson's reply was evasive (Appendix 4, pp. 162–3). He never again mentioned the slides in public.

In the light of the above, Noble's implied suggestion of a conspiracy involving the slides would have been dismissed out of hand but for a certain psychological plausibility: the specimen was doctored with ink, *ergo* the slides were also mystifications. Kammerer's suicide seemed to be an implicit admission of guilt, and the awkward evidence he had assembled in fifteen years of experimental work could be swept under the carpet. Only Przibram and MacBride stood up in public for the dead man.

Even while Kammerer was still alive, Przibram had unhesitatingly confirmed Noble on the matter of the ink, but he had rejected Noble's insinuations about the slides:

The comparison [of Kammerer's sections] with nuptial pads of other *Anura* show clearly that the callosities differ from all other known pads, resembling most those of other *Discoglossidae*, as *Bombinator*, but still more *Discoglossus pictus*. This has already been pointed out by Mr. Perkins (*Nature*, August 15, 1923, p. 238).

At the same time Przibram invited Noble to produce his own sections of *Bombinator*, for comparison with those of Kammerer's *Alytes*. Noble sent some photographs and drawings, but did not publish them. Przibram's comment on them in his last letter to *Nature*, six months after Kammerer's death, summarised the evidence:

(4) Comparing the known forms of nuptial pads in other species as to their horny spicules (Lataste, Meisenheimer, Harms, Kändler, etc.) with these [Kammerer's] drawings and photos of *Alytes*, there seems to be full specificity of these structures. Even the sections of *Bombina maxima*, the nearest approach to *Alytes*, can easily be distinguished from the photographs and drawings which Dr. Noble (Museum, New York) has sent me. The species *B. maxima* was not known to Kammerer and has never been kept alive at our Institute (see list of animals, *Zeitschrift biol. Technik u. Methodik*, 3, 163; 1913, p. 214).

(5) The histological features of Kammerer's sections of nuptial pads in *Alytes* are furthermore identical with those of a specimen found in Nature by R. Kandler (*Jenaische Zeitschrift*, 60, 175; 1924, tb. x. fig. 12) with rudimentary pads.[6]

The discovery of Kandler's specimen was a stroke of luck— though it came too late for Kammerer. The specimen Kandler found was, of course, a normal, land-breeding *Alytes*, but it was known that occasionally normal toads do develop rudiments of pads.

Both the extreme implausibility of the conspiracy charge, and the specific features of the micro-slides, point to the conclusion that Kammerer's claim to have induced nuptial pads in *Alytes* has not been refuted. Had Bateson, the Boulengers or Noble produced sections from *Bombinator* or some other toad, and shown them to be indistinguishable from Kammerer's sections, there would be room for doubt. Since Kammerer's micro-photographs are still in existence, the burden of proof rests on his critics.

3

The next point to be discussed is the black colour on the palm of the critical specimen. In 1926, this was shown to be caused by Indian ink. Two questions immediately arise: at what date was the ink injected, and what did the hand look like before the injection was made? In this context the Cambridge demonstrations in April, 1923, are crucial. Noble's report, dated three years later, says, i.a. (my italics):

> Although this specimen had presumably been carefully studied in England . . . a preliminary examination of it by me in Vienna revealed such unexpected features that Dr. H. Przibram and I have found it advisable independently to make a thorough macroscopic, histological and chemical examination of the critical features of the specimen. . . . I found the specimen to have its manus blackened both on its dorsal and ventral surfaces, the extent of the darkened area being fairly well shown in a photograph of the specimen made in Cambridge. Neither manus had the appearance of possessing nuptial pads, but both seem to have been injected with a black substance, for the blackening included some of the capillaries. An examination of the blackened areas *under moderate magnification with a binocular microscope* revealed that the colouring was not epidermal; that is, in epidermal spines, but in the derm [i.e. caused by injection].

Thus a 'preliminary' examination of the specimen 'under moderate magnification with the binocular microscope' was sufficient to make Noble realise that it had been injected with a black substance, and also to make Przibram realise that a chemical analysis was indicated.

But three years earlier the same specimen had been subjected to the same type of preliminary examination—moderate magnification under a binocular microscope—by a number of eminent biologists in Cambridge and London, some of them friendly, others sceptical or hostile, yet none of them had discovered that the 'colouring was not epidermal' but due 'to injection with a black substance.'*

* Among those present at the Cambridge demonstration were at least two expert herpetologists: Boulenger Junior and Dr. Gadow; also

Noble was aware of the importance of the Cambridge demonstration, and made a somewhat incoherent attempt to minimise it:

Dr. Przibram has brought together some distinguished names of Cambridge witnesses. With all deference to these gentlemen, I would say that the epidermis of Kammerer's specimen, which is underlaid by the black substance, appears, in part, slightly irregular. This appearance is probably due to the unequal distribution of the black substance below. At least, it required on my part the most careful manipulation of the lighting to prove that these irregularities were not in the epidermis. Further, I fail to see how anyone qualified to pronounce on the presence or absence of nuptial pads could have examined the black discolourations on the forelimbs of Kammerer's specimen without noticing their artificial character.

And yet in fact they all failed to do so. And the second crowd of biologists at the Linnean, including Bateson, again failed to do so. The only plausible explanation why they all failed to see what Noble saw quasi at first glance three years later seems to be that the doctoring was done *after* the specimen's return in its dilapidated state from Cambridge—either to preserve the aspect of the already near-vanished nuptial pads before the remaining pigment vanished altogether, or to discredit Kammerer.

The first reaction to the two reports by Noble and Przibram came from MacBride:

As I was intimately connected with Dr. Kammerer's visit to England in 1923, and as his specimens were unpacked in my laboratory and examined there before being taken to Cambridge, perhaps I may be allowed to make some comments on Dr. Noble's communication to *Nature* of August 7. As to the present condition of the *Alytes*, about which there has been so much controversy, I know nothing. Dr. Przibram's view that the specimen after its return to Vienna

Michael Perkins, J. B. S. Haldane, W. H. Thorpe, Gregory Bateson, Dr. Quastel, Dr. Harrison Matthews, Sir Sidney Harmer, Professor Stanley Gardiner, Professor G. H. F. Nuttall, G. Evelyn Hutchinson, H. N. Vevers, Professor H. Graham Cannon, Dr. Borradaile, Mrs. Onslow, F. Potts, and others. I have quoted all available eyewitness reports, without omissions.

was allowed to fade and macerate and that then a clumsy attempt at 'faked' restoration was made, appears to me probable. But this specimen was demonstrated to a continuous stream of critical observers for a whole afternoon in the Zoological Laboratory at Cambridge by Dr. Kammerer, who removed it from its case and invited examination under a lens. We all saw the spines; it was these and not the colour which convinced us. Dr. Noble may set his mind quite at rest as to the former existence of nuptial asperities.

I possess a print of the photograph which shows them—it is not a question, as Dr. Noble imagines, of two or three spines but of a whole series of minute spines regularly spaced which can be clearly seen in profile along the edge of one of the fingers. . . .

I suggest that Dr. Noble and his colleagues, instead of making aspersions on the good faith of a fellow-worker and the credulity of English scientists, would be better employed in endeavouring, as I have done, to repeat Kammerer's experiments.[7]

If the specimen had already been in the condition described by Noble when it was shown in Cambridge and London, then both groups of scientists must have been victims of collective blindness. Moreover, regardless whether the injection was done by Kammerer himself or by an assistant, Kammerer would have noticed the changed appearance of his precious specimen before taking it abroad. It is hardly conceivable that he should have embarked on an academic lecture tour knowing that the crucial object for his demonstrations was a fake, and knowing that Bateson and other hostile critics were eagerly waiting for an opportunity to prove him to be an impostor.

4

Thus the only plausible assumption is that the ink was injected *after* the return of the dilapidated specimen from England. We can now turn to the question by whom it was done and for what purpose. We cannot exclude the possibility that Kammerer did it himself, as an act of despair. His collection of preparations—the evidence of his life-work—was in ruins, and the last specimen was threatened by ruin. He might have been tempted, *knowing that the black pad had been there*, to restore its appearance artificially. He might not even have thought that he was doing something very wrong. Such sleights-of-hand are not unknown in the history of science. Several months after Kammerer's death, *Nature*, somewhat surprisingly, published a rather naive letter from an outsider, without academic qualifications, which read:[8]

> *The Nuptial Pad of Kammerer's Water-Bred Alytes*
> Some time ago, a friend of mine who was interested in my amateurish experiments on frogs, took some pictures which he intended for publication. He found it necessary to bring out some of the natural markings with ink so that they would reproduce better in print. I am wondering if the marking of Kammerer's specimens, which led to his suicide, might not have an equally simple explanation.
>
> Walter C. Kiplinger
2234 Park Avenue
Indianapolis, Ind., U.S.A.

He might have done it. Not before the journey into the lions' den in Cambridge, but later on, some time after his return. There are, however, strong arguments against this assumption. The risk of discovery was less great while the specimen stayed in the Museum, instead of being demonstrated on a lecture tour. But it was still a fatal risk to take—as events proved. There was a steady stream of visitors to the Vienna Institute, and the specimen was frequently taken out of the jar for examination by some V.I.P. Sooner or later a Dr. Noble was bound to turn up. Moreover, the injection was a staggeringly clumsy job, which Noble was able to detect almost at first sight—and Kammerer was an accomplished experimenter, whose skill in manipulation and dissection even his enemies grudgingly admitted. Van Megeeren might fake a Vermeer, but he would not use a housepainter's brush.

5

Although Noble did not mention it in his report to *Nature*, he had an assistant in carrying out the histological examination of the doctored specimen of *Alytes*. The assistant was the eminent biologist Paul Weiss, now Professor Emeritus of Rockefeller University, who, in 1926, had been administrative Director of the Vienna Institute in succession to Kammerer. He gave me the following eyewitness report of the operation which led to the discovery of the forgery (my italics):

The Rockefeller University
New York, N.Y. 10021
December 15, 1970
When Gladwyn Kingsley Noble came to the *Biologische Versuchsanstalt* to inspect and investigate the famous specimen, Hans Przibram asked me to share the task with Noble in the interest of transatlantic peace. I held then the position of 'Adjunkt', i.e., of administrative 'Assistant Director' of the Institute, which was the only official civil service position there paid by the State, and, hence, was obligated to accept the assignment, although with some hesitation. Anyhow, Noble and I got along quite well (he was only about four years older) and without ado decided to dissect the incriminated piece of skin from the pickled animal jointly and then divide the flap on the radial side, where the purported nuptial pads were to have been located, in two equal samples for histological sectioning and microscopic examination.

I remember rather distinctly the following: (1) Even in cursory surface inspection prior to dissection, the blackened area corresponded neither in topography nor in outline nor extent to my mental image of the normal morphology of other male anurans in seasonal 'heat'; but then, of course, I was no expert in the matter. (2) After blotting the water off the skin, the skin of the 'thumb' did not give the rough, non-reflecting appearance of areas with nuptial pads, but seemed as glossy as the rest. (3) *On cutting through the skin then, dark liquid spilled forth immediately, and in lifting off the skin flap, that same liquid was found to flush back and forth in the gap* between the underside of the skin and the muscles, moving freely. It had infiltrated into the skin, but whether and how deeply it had penetrated into the underlying musculature, I can't remember. I agreed with Noble that it looked like Indian ink, and I have the impression that we put some under the microscope and saw granules of the size of ink granules, but of that I am not sure. At any rate, it was dark gray, rather than black, obviously diluted by mixing freely with the fixing fluid during the preservation of the specimen. (4) As for the histological sections, I mostly remember that they were rather poor because of the rather bad state of fixation of the animal. Since the sections were made transversally (at right angles to the surface), no census of pigment cell density to test for local accumulation could, of course, be made. I don't recall to have noted any signs of the serrated cuticular appearance of the skin of nuptial pads; in view of the bad fixation, these might have been rubbed off during previous manipulations.

In conclusion, I personally have little doubt that the 'pigment', presented as diagnostic, was man-made artefact in substance and localisation. A second specimen which Przibram and I discovered some time later in one of our museum jars showed an equally aberrant 'pigment' patch under the skin, but I did not investigate it further, nor was I asked to, and I have no idea what has happened to it.

6

As the bits of evidence kept accumulating, I felt, to quote Bateson, 'a strong curiosity' to see a midwife toad with nuptial pads faked by injection of Indian ink. But the doctoring would have to be done by an expert, convincing enough to deceive a bevy of English and other naturalists. I happened to know Professor Holger Hydén of the University of Gothenburg, one of the leading experimental cytologists in Europe—the first who succeeded in extracting, free-handed, the nuclei of nerve-cells.* He thought it would be amusing to try his hand at forgery, using advanced surgical techniques, to give the forger a fair chance.

There are no *Alytes* to be found in Scandinavia, but Hydén got hold of two specimens, a male and a female, from a dealer in Milan. His first 'progress report' is dated October 27, 1970:

On October 26
the left and right hand, palmar view, photographed in colour, Kodak, Hasselblad camera. [One of these photographs is shown on the jacket cover.]

---

* Hydén is the founder of the theory that memory is based on biochemical changes in brain-cells on the molecular level, affecting RNA and specific proteins.

the left hand of both *Alytes* injected with Pelikan Indian ink, syringe, hypodermic 20.

Penetrated skin on ventral side at the wrist up to the base of the first finger and injected slightly, retracted the syringe and injected again towards and into the thumb-ball. The Indian ink spread easily on the ventral and also slightly over the dorsal side of the left edge of the left hand and downwards on the radial aspect of the arm. Washed and replaced in solutions. The view very realistic. Photographed both as above.

(Hydén preserved one specimen in alcohol, the other in formaldehyde, because we were not sure which was used by the Vienna Institute.)

The next communication, dated November 6, included the first photographs of the forged specimens:

Here are the colour photos. It is easy to see in the *Alytes* fixed in alcohol (♀) that we are dealing with artefacts. The Indian ink is clumping together in irregular lines:

and I doubt that the *Alytes* of Kammerer was fixed in alcohol.

If you look at the ♂ *Alytes* [in formaldehyde], there is *quite* an attractive view, at least in the palmar area. Perhaps the dark dots [of ink] near the wrist are suspicious, I don't know:

November 16

Enclosed the new photographs taken on Nov. 13 of the right hand of the formaldehyde-fixed male, injected [the second time] in the morning of Nov. 13 with Indian ink (massaged lightly the fingers and wrist to get a good filling).

Today, the left hand of the same male was cut off and the digital part taken for embedding, sectioning and staining.

November 30

So far as I can see every trace of Indian ink inside the tissue has been washed away by the fixing and staining solutions . . .

Here the progress reports stop, because in December 1970 Professor Hydén visited London. He brought the two doctored specimens with him, and any expert in herpetology is welcome to inspect them. The gist of his report is as follows. The specimen preserved in *alcohol* looked at first (October 27) 'very realistic'. But by November 6 'it was easy to see that it was an artefact. If you looked at it under the microscope, you saw small spherical or elongated clumps under the skin. And these looked to me clearly as artefacts.' He then injected it again on November 13, 'and that worked now the second time much better, for some reason'. But soon there was too much blackening, and the whole hand looked as if dipped in ink. Then it again faded to grey, except near the ridges.

The *formaldehyde* specimen, too, looked at first 'very realistic', and on November 6 still presented 'quite an attractive view'. But a few days later 'it was a little bleached and faded, and when I gently massaged the arm and the hand I could see that the fluid below the skin was moving. And since the Pelikan Indian ink is easily soluble in water, it was quite clear what had happened. It was dissolving in the formaldehyde solution and the tissue fluid.' After the second injection, on November 17, the specimen 'looked really fine to begin with'. But a week later 'it had sort of bleached out again. Evidently, the ink had dissolved a little more rapidly than the first injection of October 27.' He made a third injection on November 30 and observed that 'the blackening of the palmar pad of the hand more or less disappeared in a few days'.

In January, 1971, Hydén tried alternative techniques to stabilise the ink and prevent the bleaching of the black spots. He was unable to obtain more *Alytes*, so he used ordinary frogs (*Rana bombina*). He tried a mixture of ink and glycerol, but that, too, dissolved readily in alcohol or water. He tried paraffin oil, but that did not take up the ink. The best results were obtained with a mixture of ink and warm gelatine. This time only formaldehyde was used as a preservative, because it gave the better results—the specimens in alcohol shrunk in size. Before the injections were made the specimens were kept at room temperature.

Two months later, by the middle of March, there was no perceptible fading of the black colouration. This spoke in favour of the gel method. The blackened area 'covered the palmar side, including the thumb ball and the basal part of the first finger, spreading over the lateral side of the fourth finger'. But now there was another drawback. This black patch had sharply circumscribed boundaries which corresponded to the area of the congealed gelatine underneath; it did not show the natural transition from black to grey towards the fingers, but a smooth, uniform, glossy blackness with little natural pigmentation. It looked altogether artificial.

A further point emerged from the gel-experiment. Hydén made microtome sections across the injected area.* 'The tissue was well preserved. There was a blackish substance with India ink granules between bundles of muscles and vessels, but no such substance was visible within capillaries or vessels [as in Noble's report]. The gelatine with its content of Indian ink had precipitated in the formaldehyde.' Thus the black substance did not dissolve in the dissecting fluid as the diluted ink had done. After injection, the warm gelatine mass had congealed in the tissue, and on dissection could not, of course, spill forth, as described in the reports by Noble and Paul Weiss.

I have left out the technical refinements in Professor Hydén's account. The following conclusions emerge:

* Seven-micron sections stained with haematoxylin and eosin.

The Cambridge and Linnean demonstrations took place in 1923. Noble's examination took place in 1926. If the injection was made with ink alone, it seems highly unlikely that its effects would have survived three years without fading through dilution. The use of a gelatinous solution might have preserved the black patches, but for the reasons just mentioned this method can be fairly safely ruled out. The most likely hypothesis, therefore, is that ink alone was used and, in view of its rapid fading, that the injection was made shortly before Noble's visit.

7

Needless to say, the experiments cannot be regarded as conclusive, and were not intended to be so. Hydén does not exclude the possibility that a different method could be invented which would produce more convincing results. But in the light of the supporting evidence it seems highly probable that the forgery was committed *after Kammerer's lectures in England and only a few days before Noble's expected visit.* But who did it?

I have discussed a number of reasons why it seems improbable that Kammerer himself did. To these must be added the fact that from the end of 1922 onward Kammerer no longer worked at the Institute. On December 1, 1922, he was granted one year's leave of absence, and on October 30, 1923, his final retirement was agreed on. Most of these last three years he spent on lecture tours or in the Soviet Union. We do not know whether, after his retirement, he still paid occasional visits to the Vivarium, but if so, he was now in the position of a distinguished visitor, to whom attention would be paid, and it is not easy to see how he could have removed the specimen from its museum shelf and carried out the operation, unnoticed—unless he wore a black mask.

We now come to the hypothesis that the forgery was done by a technician or member of the staff. One might speculate that it was some adoring female lab assistant, desperately anxious to help the dear, sad Professor. Her heart would have been overflowing with love, and the syringe with ink—she would probably have been quite ignorant of the difference between natural pigment and Indian ink, and have thought the more the better. This would somehow fit the ambiance of romantic operetta, which kept intruding into Kammerer's life; and Przibram's first reaction to Noble's disclosure was, as we have seen, on these lines.

But on reflection, he changed his mind regarding the motives of the forgery. In his comments to Noble's paper he had suggested that 'someone has tried to preserve the aspect of such black nuptial pads in fear of their vanishing by the destruction of the melanin through exposure to the sun in the museum case'.[9] But a few weeks later, after Kammerer's suicide, Przibram mentioned in an obituary article[10] that in 1918 'a madly jealous colleague had falsely claimed to have refuted the colour changes in Kammerer's salamanders in the very first generation,* and that later on this man had to be 'temporarily locked up in a mental asylum'. The term 'colleague' (*Kollege*) indicates somebody working with Kammerer at the Institute. Who he was, and what became of him, we do not know, owing to Hans Przibram's discretion—which, in that gossipy town, was proverbial. He even went out of his way to protect a former assistant—a certain Dr. Megusar—against 'any suspicion of having made the injection' since he 'was killed at the Wolhynian frogt on August 3, 1916'.[11] And Przibram even gives the reference for the obituary '(*Archiv für Entwickl-Mech.*, 42, 222: 1917)'—presumably as proof that the man was really dead.

After Przibram's death, his brother, Professor Karl Przibram, wrote:

> He [Hans Przibram] remained convinced of the genuineness of Kammerer's observations, and repeatedly said in private conversations that he thought he knew who committed the forgery to discredit Kammerer, but could not make a public statement for lack of sufficient evidence.[12]

In two letters to me, Professor Karl Przibram repeatedly emphasised the same point:

> My brother always remained convinced of the honesty of Kammerer, and said on several occasions that he believed he knew the identity of the forger, but could not publish it for lack of sufficient evidence. If I may add my own view, I regarded Kammerer as far too intelligent to commit such a clumsy blunder.[13]

* These adaptive changes in the first generation were not in dispute—only their inheritability.

My brother certainly believed in Kammerer's innocence, and was convinced that the sole purpose of the forgery was to discredit Kammerer and his work.[14]

Was Przibram's suspect the same mad colleague who, having tried to refute the salamander experiments in 1918, tried to destroy the *Alytes* evidence in 1926? We do not know. But whoever he was, he knew that the visit of an American reptile expert was impending, and that this expert was hostile to Kammerer. This provided an ideal opportunity—and a powerful incentive—for injecting the ink some time before Noble's arrival.

The motives of Przibram's suspect may have been personal jealousy, or even political. The Soviet film *Salamandra*, which I mentioned earlier on, assumed that the forger acted for political motives (see Appendix 2): evidence for the inheritance of acquired characters would be a mortal blow to the racialist doctrine of the genetically determined excellence or inferiority of nations. The hypothesis is not quite as far-fetched as it sounds: a fanatical Nazi—perhaps the 'mad colleague'—may well have been tempted to carry out such a lunatic act. Austria in the middle twenties was steadily moving toward civil war, to the tunes of *The Merry Widow* and 'Yes, We Have No Bananas'. Political assassinations were increasing in frequency; Socialists and Nationalists each had their private armies— *Schutzbund* and *Heimwehr*; and the *Hackenkreuzler*, the swastika-wearers, as the Nazis of the early days were called, were growing in power. One of the centres of ferment was the University of Vienna where, on the traditional Saturday morning student parades, bloody battles were fought. Kammerer was known by his public lectures and newspaper articles as an ardent pacifist and Socialist; it was also known that he was going to build an institute in Soviet Russia. An act of sabotage in the laboratory would have been just as much in keeping with the climate of those days as an act of misplaced devotion by some lovelorn Fräulein armed with a syringe.

8

Rumour and conjecture were rife in Vienna after Kammerer's death, but one curious fact seems to have been overlooked. Noble's disclosures were published in *Nature* on August 7,

1926, but his investigations were carried out at least six months earlier, and their results would have spread through the academic grapevine. And yet a few days before his suicide—that is, around September 20—Kammerer visited the Soviet Legation in Vienna and 'with much zest gave instructions regarding the crating and transport of the scientific apparatus and machines which he had ordered for his future experimental institute in Moscow'.[15] He was to start work in Russia on October 1. In other words, the Russian Academy did not revoke their invitation after the forgery had been discovered, although the discovery was made half a year earlier and was published six weeks earlier, so that they had had plenty of time to back out. Even if a written contract had been signed, the Russians could easily have found pretexts for delay and prevarication, instead of providing the means for Kammerer to buy the equipment and take charge of the crating in the premises of the Legation. Apparently their trust in him was unshaken, and we may assume that they had good reasons for it. After all, Pavlov's Institute enjoyed a world-wide reputation, and could not appoint a man to take charge of a department and build a new experimental station, if they thought him an impostor. The Russians were not deterred by the scandal; it was Kammerer who wrote in one of his farewell letters, addressed to the Soviet Academy, that, although he had no share in the forgery, he no longer considered himself qualified to accept the post offered to him.

What made the Russians so sure? It is possible that they had their own sources of information within the Vienna Institute—a Party member or sympathiser. It would have been contrary to tradition if they had appointed a foreigner to a very responsible position without compiling a dossier of intelligence reports, including, as a matter of routine, information on his political views, friends, private life and so on. The accusation of fraud must, of course, have been gone into. It is not impossible that the report included some inside information which put the minds of those responsible for the appointment at rest. This, of course, is pure conjecture. But it is difficult to find an explanation for the notoriously suspicious Russians' faith in Kammerer, other than that they had good reason to believe that it was somebody else who had injected the ink.

Ten

In that syrupy operetta *Das Dreimäderlhaus*, Franz Schubert
pays court to all three maiden daughters of his patron. Paul
Kammerer broke the record by falling in love successively with
the five famous Wiesenthal sisters. Four of them were ballet
dancers, the fifth played the violin. The eldest, Grete Wiesenthal, was solo dancer and Ballet Master of the Vienna Opera;
later she had her own school where she taught a new style of
dancing—the Wiesenthal Style. She also had one of the last
literary salons in Vienna.

'I think Father was in love with every one of them', Lacerta
says. It apparently took him about ten years to decide, by a process of elimination, that it was Grete whom he really meant.

At least two of the romances were, on Lacerta's evidence,
purely platonic. When he was thirty, he composed a dance
dedicated to *Bertha*, one of the younger ones; it is called 'The
Wiesenthal Ländler'. But there is a motto on top of the score,
a quotation from the poet Peter Altenberg, who belonged to their
circle: '*Du mit Deiner süssen, merkwürdigen Schwester, Bertha.*'
This makes the dedication somewhat ambiguous, and leaves one
in doubt who is meant by the *Du* who has the 'sweet, strange
sister'. Anyhow, Kammerer later spoke of Bertha as *die Eis-
jungfrau*, the Icy Virgin; so that was a flop. *Elsa* apparently too;
she married the well-known painter Rudolf Huber. He painted
portraits of Kammerer and Felicitas; the Kammerers, Hubers
and the other four sisters 'met regularly for afternoon teas and
took turns in acting as hosts'. Then the youngest, *Martha*, also

married, 'and it was during her married life that my father became infatuated with her'. Impoverished as he was, he made her extravagant gifts. About *Hilda* (the one who played the violin) we are told nothing.

'Grete was the eldest and at the same time the last to come into the picture.' When she did, the comedy took a tragic turn.

Grete Wiesenthal was then forty, and at the height of her fame. She was married to the Swedish architect Somerskjöld, who does not come into the picture. To what extent she reciprocated Kammerer's feelings, we do not know. Lacerta says that 'he was very much in love with her, but her response did not go far enough one way or another'. It went, however, far enough to make their liaison widely known in Vienna. Some of the obituaries hinted that the cause of his suicide was her refusal to accompany him to Moscow.

Yet in view of the emotionally unbalanced state in which he lived towards the end, it would be futile to look for a single cause. As far as we know, he wrote four farewell letters (apart from the note 'to the person who will find my body'), and in each of them he seems to have given different reasons for his suicide.

One of the farewell letters was addressed to the Moscow Academy of Science. It refers to Noble's disclosures and continues: 'In view of these facts, though I had no share in the faking of the specimen, I can no longer consider myself fit to accept your invitation. But I also find it impossible to accept this denial of my life-work, and hope that I shall have the courage and force to put an end to my failed life tomorrow.'[1]

Another letter was addressed to Felicitas. Only a short extract is known of its contents, which, in its published version, states that he 'finds it impossible to accept the Russian offer. His ties with Vienna are too strong and the only way out of this dilemma is to end his life'.[2]

A third letter was addressed to his intimate friend, Baron Willy von Gutmann. 'It repeats, in different words, the reasons given to the Moscow Academy, and adds a second reason of a personal and private nature which has nothing to do with scientific considerations.'[3]

The fourth letter was addressed to Grete Wiesenthal. Its contents are unknown. It arrived while the sender was still alive, walking on the Schneeberg.

The most informative comment appeared, two days after his death, in the *Neue Freie Presse*:

DR. KAMMERER'S SUICIDE

*Communications Received From Sources
Near To Him Concerning The Causes Of
His Deed*

During his very last days Professor Kammerer still bought and ordered equipment on a large scale for his Moscow Institute, and it had been arranged that the removal men would come in on Thursday, the day on which he died, to crate his books and furniture for transportation to Moscow. He had been completely absorbed in his plans for the move, and even in the last days had invited one of his friends, a Viennese scientist, to collaborate with him in Moscow on an important research project in the study of heredity. The subject of the study had been thoroughly discussed, and the experimental procedure worked out in detail. Then, all of a sudden, he expressed the intention of staying only for a short time in Moscow. He hoped that the work he wanted to do in Moscow would secure him an invitation to a German institute for genetics, and he seemed to have good reasons to believe that one of the new German universities would offer him a position.

However, the fatal decision to take his own life seems to have been determined by the fact that a Viennese artiste who was close to his heart was unable to make up her mind to follow him to Moscow, and the cyclic depression which had once already, three years earlier, driven him to attempt suicide by veronal seems suddenly to have gained the upper hand.[4]

The conclusion seems to be that he had hoped to the last minute that Grete Wiesenthal would consent to follow him to Moscow. That hope may have been founded on an illusion; but nevertheless her refusal was the last straw. On that same Thursday on which the removal men were busy packing up his furniture, he was walking on the mountain until he found the courage to do it.

There is a macabre detail that must be mentioned. The day after his death the Viennese evening papers reported:

About the circumstances of the final act, the following details have come to light. Dr. Kammerer arrived in Puchberg on Wednesday evening and spent the night in an inn, *The Rose*. On Thursday morning he went for a walk from which he did not return. He took a narrow footpath which leads from Puchberg past the Theresa Rock to Himberg. At the Theresa Rock he sat down on the roadside and carried out his deed.

Dr. Kammerer was found at 2 p.m., by a roadworker from Puchberg, in a sitting position. He was leaning with his back against the Rock and his right hand was still holding the revolver. In spite of the fact that he held the weapon in his right hand, the bullet had entered the skull from the left side, above the ear. It traversed the head and emerged through the right temple. The impact also damaged the right eye. Death must have been instantaneous.[5]

Another paper commented:

Apparently Dr. Kammerer committed suicide in a rather complicated way. He held the gun in his right hand, whereas the bullet . . . etc.

It was a difficult feat to achieve, and it carried the risk of botching the job. Anybody with even a slight knowledge of the anatomy of the brain must be aware of that. With one's arm across one's face it is difficult to control the angle of the weapon, even in front of a mirror. And the wrong angle might mean blindness or a crippled life, instead of death.

I asked friends in the medical profession whether they had come across similar cases, but they had not. The only explanation of sorts was offered, not by a psychiatrist, but by an intuitive woman. Though Kammerer was an abstemious man, he might have had a few drinks to give himself Dutch courage before he set out on his last walk. Sitting with his back to the rock, he might have hesitated for a long time, then, with a sudden swinging gesture, brought his arm around his face and the gun to the left ear. There is more panache in such a gesture than in the conventional lifting of the gun, right hand to right temple. Even if not drunk on alcohol, the gesture would fit an impulse of sudden desperate exaltation—and end with a flourish. His death had a touch of melodrama, but so had his life. He was a Byron among the toads.

2

William Bateson did not live to see Kammerer's undoing. He died in February, 1926, at the age of sixty-four, about the time when Noble discovered the evidence that the pads on the last specimen of *Alytes* were faked.

It was a Pyrrhic victory for still other, deeper reasons. By 1924, Bateson had come to realise, and told his son in confidence, 'that it was a mistake to have committed his life to Mendelism, that this was a blind alley which would not throw any light on the differentiation of species, nor on evolution in general'.[6]

Hans Przibram's last years are painful to relate. This kind, somewhat pedantic and unworldly man seemed to have been singled out to live through a succession of catastrophes which would seem extraordinary, if they had not been symbolic for that age. Przibram was a Jew, but he too, like Kammerer, had married an aristocrat, from an old Polish family, Countess Komarovska. This kind of alliance was not unusual in the old monarchy, where prominent scientists had a much higher social standing than in the West; actually, the Countess had previously been married to the Russian Prince Galitzin. She was a romantic, highly strung woman who, in 1933, committed suicide by taking veronal and slashing her wrists. They had three daughters.

In 1935 Przibram married again—this time a Jewish widow. He was completely unaware of the danger that Hitler represented to Austria. His former assistant, Paul Weiss, then at the University of Chicago, offered to find him a position in America. Przibram refused; he would not believe that Austria could sink into barbarity.

When the Nazis annexed Austria in 1938, the husband of Marguerite, the eldest Przibram daughter, committed suicide—but he lingered on for a week before he died. She had a nervous breakdown and had to be put into a psychiatric hospital. During the war, she and the other inmates of the hospital were deported to Minsk, from where nobody returned.

The husband of the second daughter, Vera, was killed by the

Nazis. She managed to get away, married again—a Hungarian count—and now lives in Canada. The third daughter, Doris, also escaped, and is married to an Austrian diplomat.

Just before the outbreak of war, Przibram was at last persuaded to emigrate with his wife to Amsterdam. Forever hopeful, he expected to return shortly to Vienna; in the meantime, he continued his work, in a Dutch laboratory, on the chemistry of the pineal gland. When the Nazis occupied Holland he continued his work until, in 1943, he and his wife were deported to the concentration and extermination camp Terezin, formerly Theresienstadt. (It is in Bohemia, fifty miles due north from a town called Pribram.) The last direct news from him was a postcard, dated Amsterdam, April 21, 1943, on which he had hurriedly scribbled in pencil: 'We have been invited to travel to Theresienstadt . . .'[7]

He had probably thrown it out of the cattle-truck, and some friendly Dutchman had posted it.

In the spring of 1944 news came that Hans Przibram, aged seventy, had died in Theresienstadt. The manner of his death was not indicated. His wife somehow managed to get hold of poison, and killed herself.

His beloved Vivarium, the Sorcerers' Institute, went up in flames during the Russian bombardment of Vienna. The Austrian Academy sold the ruins to a show-business operator.

Epilogue

I

The following is a brief summary of the tangled case of the midwife toad.

Did Kammerer breed water-mating *Alytes* with hereditary nuptial pads, and did the critical specimen show the pads before it was tampered with?

The main arguments in favour of a positive answer are the microscope photographs, still in existence, showing features which differ from sections of pads from other species, but are apparently similar to the rudimentary pads of Kändler's specimen found in the wild. These seem to exclude the possibility that sections from another species were substituted for those from *Alytes*, quite apart from the inherent improbability of a conspiracy involving Kammerer, Przibram and the histologist, Miss Kermauner.

Supporting evidence is provided by the testimonies of a host of biologists who saw the specimen exhibited in Vienna, Cambridge and London, and by the photographs of it taken by Congdon (1919), Reiffenstein (1922) and B. Stewart in Cambridge (1923). Although some biologists who examined the specimen with microscope or lens remained sceptical, doubting whether the asperities and spines were sufficiently pronounced to be identified as nuptial pads, nobody spotted traces of Indian ink which, three years later, were obvious to Noble, Przibram and P. Weiss under 'moderate magnification', and are equally obvious in the specimens injected by Professor Hydén.

An additional argument in favour of the authenticity of the

preparation exhibited before and during the English visit is the grave risk of exposure which Kammerer would have run in submitting to critical examination a forged specimen, regardless whether the forgery was committed by himself or somebody else.

All this points to the conclusion that the ink was injected, after the return of the specimen to Vienna, by a member of the Institute's staff, with intent to discredit Kammerer. This is the opinion which the Head of the Institute repeatedly expressed in public and in private. The announcement of Dr. Noble's forthcoming visit may have decided the timing of the injection, which, in the light of Hydén's experiments, was probably done a short while before Noble's arrival. As Kammerer no longer worked at the Institute, he would have been unaware that somebody had tampered with the specimen and thus he willingly consented to the histological examination.

The possibility cannot be entirely excluded that he did the injecting himself in a moment of despair. He could not get over the ruin of his collection, and in the unstable state of mind of his last year might not even have thought that to 'restore' the pad which had once been there would be a crime. The history of science abounds in examples of correcting nature in a good cause.

Against this assumption speak the clumsiness of the forgery, the persistent risk of discovery and, on the psychological side, that transparent sincerity of manner to which even his opponents testified.

My personal belief that he did not do it is based as much on the impression I formed of the man's character as on the supporting evidence. I did not have that belief when I started on this essay. Nobody who reads about Kammerer in current books on biology could believe in his innocence. But as the source material came in from the archives, and the eyewitness reports from surviving participants in the drama, I realised that the accounts in these books were distorted, based on hearsay long after the event, and had hardly any relation to the facts (see Appendix 2, 'The Legend') I did not start with the intention to rehabilitate Paul Kammerer; but I ended up with an attempt to do so.

In his lifetime he was the victim of a campaign of defamation by the defenders of the new orthodoxy—a situation which recurs with depressing monotony in the history of science. His opponents refused to admit that through his breeding experiments he had made out a *prima facie* case, but were either unable or unwilling to repeat them. After his death in dismal circumstances, they felt freed from this obligation. The ink, injected by an unknown hand, was remembered; fifteen years of experimental work with lizards, salamanders, sea-siphons and toads could now be conveniently forgotten, and with it the challenge it represented to the orthodox position. The skeleton was safely locked in the cupboard.

His most powerful and determined opponent was one of the founders of that orthodoxy, William Bateson. In looking back at the controversy which ranged over fifteen years, one is struck by Bateson's strategic skill and Kammerer's lack of it. Kammerer stumbled on the pads of his water-bred *Alytes* by pure chance in an experimental series with a quite different purpose: to induce a change in the toads' mating habits. He originally attached little importance to the pads, and throughout the controversy he kept repeating that they were 'by no means a conclusive proof of the inheritance of acquired characters'.[1] Bateson, on the other hand, ignored the experiments with *Ciona* (which Kammerer regarded as the most important) and with all other species, and concentrated his attack first on *Salamandra* and *Alytes*; then on the pads of *Alytes* alone. Subsequently he again ignored the microtome sections across the pad, and concentrated only on the *position* of the pad. Lastly, he ignored the presence of the pad on the back of the hand—the 'correct position', and concentrated on the black mark on the palm, maintaining that it was in the 'wrong place', and ignoring all arguments to the contrary. Thus the battlefield was of Bateson's choosing, and he managed to narrow it down more and more until it became a trap. The debate on the origin of species was reduced to the case of the nuptial pads of the midwife toad.

2

Assuming that Kammerer's experiments were repeated and confirmed—what would they prove?

They would certainly not prove that Lamarckian inheritance is the governing principle of evolution. Some leading physicists are opposed to the orthodox theories in contemporary quantum-mechanics, but that does not mean that they want to go back to the physics of Aristotle. If Darwin was wrong in some important respects, that does not mean that Lamarck was right. But it might mean that Lamarck was not completely and entirely wrong. And it is conceivable that the type of experiment which was Kammerer's speciality might just fit the tiny gaps in the 'Weismann barrier' which prevents acquired features from interfering with the blueprint for future generations.

How this could be done is difficult to visualise in terms of contemporary biochemistry. On the other hand, there is today again a growing conviction among biologists that Darwinism alone cannot explain the evolution of species. Darwin, we remember (p. 32), was the first to realise this—so he fell into the Lamarckian heresy. Bateson and others followed in his footsteps. Then came Mendel. 'Only those', Bateson wrote in 1924, 'who remember the utter darkness before the Mendelian dawn can appreciate what happened.'[2] But a few lines further down he went on: 'Though Mendelian analysis has done all this, it has not given us the origin of species.'[3] This was written two years before his death, at the time when he realised that it had been 'a mistake to commit his life to Mendelism' (p. 121). But even ten years earlier his disillusionment with the Darwinian theory of natural selection was almost as complete as his detestation of Lamarckism:

> The many converging lines of evidence point so clearly to the central fact of the origin of the forms of life by an evolutionary process that we are compelled to accept this deduction, but as to almost all the essential features . . . we have to confess an ignorance nearly total. The transformation of masses of population by imperceptible steps guided by selection is, as most of us now see, so inapplicable to the facts, whether of variation or of specificity, that we can only marvel both at the want of penetration displayed by the advocates of such a proposition, and at the forensic skill by which it was made to appear acceptable even for a time.[4]

Bateson coined the term 'genetics'; W. Johannsen coined the term 'gene'. He was another pioneer of neo-Darwinism on Mendelian lines. By 1923 he too realised that all mutations ever induced in the fruit-fly—the geneticist's favourite experimental object—had been either deleterious or trivial; and that to regard chance mutations in the genes as an explanation of the evolutionary process was a highly improbable speculation, not supported by any empirical evidence:

Is the whole of Mendelism perhaps nothing but an establishment of very many chromosomal irregularities, disturbances or diseases of enormously practical and theoretical importance but without deeper value for an understanding of the 'normal' constitution of natural biotypes? The Problem of Species, Evolution, does not seem to be approached seriously through Mendelism nor through the related modern experiences in mutations.[5]

Then, in the 1950s, came another dawn: the discovery by Crick and Watson of the chemical structure of DNA, the nucleic acid in the chromosomes, carrier of the 'hereditary blueprint'. The Weismann doctrine, that nothing* that happens to an organism in its lifetime can alter that blueprint, was now elevated into Crick's so-called 'Central Dogma', which states that 'information can flow from nucleic acids to proteins [i.e. from blueprint to building block] but cannot flow from protein to nucleic acid'. But dogmas are brittle structures. On June 25, 1970, the *New Scientist* announced: 'Biology's Central Dogma Turned Topsyturvy.' *The Times* Science Report followed suit: 'Big Reverse for Dogma of Biology.' The report concluded:

It is too early to say what consequences may follow from the demonstration that DNA can be copied from RNA,† but at the least the central dogma now seems to be an oversimplification.[6]

What happened was that three separate cancer-research teams at M.I.T., Wisconsin and Columbia Universities had published papers with experimental proof that certain viruses which cause cancer in animals, once they invade the host cell, can produce their own hereditary DNA. Howard Temin of Wisconsin had predicted this result some seven years earlier. But because it contradicted the 'Central Dogma', and smacked of the Lamarckian heresy, 'Teminism' was largely ignored, until D. Baltimore of M.I.T. and Sol Spiegelman, Head of Columbia's Institute for Cancer Research, confirmed Temin's claims. *Plus ça change* . . .

* Short of destructive catastrophes, of course.
† RNA is the 'messenger' substance which transmits the instructions of the DNA blueprint to the protein factories in the cell. According to the 'Central Dogma', information can pass only in one direction: DNA to RNA to protein.

It would, of course, be silly to jump to the conclusion that because viruses can produce hereditary changes in a cell, therefore continued piano practice by the parents will make them beget musical prodigies. All that these discoveries (and some related ones) indicate is that the 'Weismann barrier' is not as absolute as the dogmatic view would have it.

Within roughly the same period, other important papers* were published in *Nature*, which some biologists consider to be the beginning of the end of neo-Darwinism in its present form. The subject is too technical to go into—except for quoting the opening paragraph of Salisbury's article:

> Modern biology is faced with two ideas which seem to me to be quite incompatible with each other. One is the concept of evolution by natural selection of adaptive genes that are originally produced by random mutations. The other is the concept of the gene as part of a molecule of DNA, each gene being unique [specific] in the order of arrangement of its nucleotides. If life really depends on each gene being as unique as it appears to be, then it is too unique to come into being by chance mutations. There will be nothing for natural selection to act on.

The arguments he uses to show that random mutation and natural selection alone could not have kept evolution going, without some additional principle being involved, are derived from biochemistry and modern information theory. Yet essentially they are only a sophisticated way of formulating Waddington's remark, which I have quoted before (p. 30), that it is not reasonable to try to build a habitable house by throwing bricks together in a random heap. Salisbury concludes: 'In the evolution of life on Earth, we are dealing with millions of different life forms, each based on many genes. Yet the mutational mechanism as presently imagined could fall short by hundreds of orders of magnitude of producing, in a mere four billion years, even a single required gene.'

Professor W. H. Thorpe summed up the present situation when he wrote of 'an undercurrent of thought in the minds of perhaps hundreds of biologists over the last twenty-five years'[6a] rejecting the neo-Darwinist orthodoxy. One of its most persis-

* Salisbury, F. B., *Nature, 224,* 342 (1969).
  Smith, J. M., *Nature, 225,* 563 (1970).
  Spetner, L. M., *Nature, 226,* 948 (1970).

tent critics has been the veteran biologist, Ludwig von Bertalanffy:

> I think the fact that a theory so vague, so insufficiently verifiable and so far from the criteria otherwise applied in 'hard' science, has become a dogma, can only be explained on sociological grounds. Society and science have been so steeped in the ideas of mechanism, utilitarianism and the economic concept of free competition, that instead of God, Selection was enthroned as ultimate reality. On the other hand, it seems symptomatic that the present discontent with the state of the world is also felt in evolution theory. I believe this is the explanation why leading evolutionists like J. Huxley and Dobzhansky (1967) discover sympathy with the somewhat muddy mysticism of Teilhard de Chardin. If differential reproduction and selective advantage are the only directive factors of evolution, it is hard to see why evolution has ever progressed beyond the rabbit, the herring, or even the bacterium which are unsurpassed in their reproductive capacities.[7]

3

Only a fool or a fanatic could deny the revolutionary impact of Darwinism on our outlook. If I have concentrated on its shortcomings, it is partly because of that philosophical bias to which von Bertalanffy alluded; and partly for reasons explained in the opening pages of this essay. The totalitarian claim of the neo-Darwinists that evolution is 'nothing but' chance mutation plus selection has, I think, been finally defeated, and a decade or two from now biologists—and philosophers—may well wonder what sort of benightedness it was that held their elders in its thrall. Darwinian selection operating on chance mutations is doubtless a part of the evolutionary picture, but it cannot be the whole picture, and probably not even a very important part of it. There must be other principles and forces at work on the vast canvas of evolutionary phenomena.

In fact, quite a number of such principles have been proposed, and some of them experimentally demonstrated, by various eminent biologists. In a previous book* I have assembled some of these separate theories and tried to fit them to-

* *The Ghost in the Machine*, Part Two: Becoming, p. 115 *et seq.*

gether. I shall not discuss them again, but just mention some alternatives, or corollaries, to Darwinian theory which have been proposed. There is the so-called 'Baldwin effect', which was rediscovered independently (as Mendel's paper was) by C. H. Waddington and Sir Alister Hardy. Waddington revived it in his theory of 'genetic assimilation' and Hardy in his own theory of 'behaviour as a selective force'. More recently we had Teminism, Salisbury's critique, and further proofs of cytoplasmic inheritance (e.g. Sonneborn, 1970). Much discussed in their time were also Garstang's 'paedomorphism', L. L. Whyte's 'internal selection', and much further back, Geoffroy de St. Hilaire's *'loi du balancement'*. The purpose of this eclectic list minus explanatory comments is merely to indicate that all sorts of corrections and amendments to Darwinism have been proposed by biologists over the years with varying degrees of plausibility, and that the naive version of Lamarckism current in Darwin's own day is not the only alternative. There seems to be every reason to believe that evolution is the combined result of a whole range of causative factors, some known, others dimly guessed, yet others so far completely unknown. And I do not think one is justified in excluding the possibility that within that wide range of causative factors a modest niche might be found for a kind of modified 'Mini-Lamarckism' as an explanation for some limited and rare evolutionary phenomena.

They must, by necessity, be rare, for a simple reason, best explained by an analogy. Our main sense organs for sight and hearing are like narrow slits which admit only a very limited frequency range of electro-magnetic and sound waves. But even that reduced input is too much. Life would be impossible if we were to attend to each of the millions of stimuli which constantly bombard our receptor organs in a 'blooming, buzzing confusion'—as William James called it. Thus the brain and the nervous system function as a hierarchy of filtering and classifying devices which eliminate a large proportion of the input as irrelevant 'noise', and process the relevant information into a presentable shape before it is admitted to consciousness. A typical example of this filtering process is the so-called 'cocktail-party phenomenon'—our ability to isolate a single voice from the general buzz.

Now what the 'Weismann barrier' or 'Central Dogma' really means is that a similar filtering apparatus must protect the hereditary substance against the blooming, buzzing confusion of biochemical incursions, which would otherwise play havoc with the continuity and stability of the species. If every experience of the ancestors left its hereditary imprint on the progeny, the result would be a chaos of forms and a bedlam of instincts. But that does not exclude the possibility that some 'cocktail-party phenomenon' might, in rare cases, be present in the evolutionary process. That is to say, that the 'Weismann barrier' might not be an impenetrable wall, but a very fine-meshed filter, which can only be penetrated under special circumstances.

Some classical examples quoted over and again in the literature seem almost to cry out for a 'Mini-Lamarckian' explanation:

> There is, for example, the hoary problem why the skin on the soles of our feet is so much thicker than elsewhere. If the thickening occurred *after* birth, as a result of stress, wear and tear, there would be no problem. But the skin of the sole is already thickened *in the embryo* which has never walked, bare-foot or otherwise. A similar, even more striking phenomenon are the callosities on the African warthog's wrists and forelegs, on which the animal leans while feeding; on the knees of camels; and, oddest of all, the two bulbous thickenings on the ostrich's undercarriage, one fore, one aft, on which that ungainly bird squats. All these callosities make their appearance, as the skin on our feet does, in the embryo. They are inherited characters. But is it conceivable that these callosities should have evolved by chance mutations just exactly where the animal needed them? Or must we assume that there is a causal, Lamarckian connection between the animal's needs and the mutation which provides them?[8]

It is admittedly difficult to see how an acquired callosity could conceivably produce a change in the chromosomes. But, as Waddington has pointed out, 'even if improbable, such processes would not be theoretically inexplicable. It must be for experiment to decide whether they happen or not.'[9] Waddington has even produced a 'speculative model' to show a possible way how changes in the activities of body-cells could affect the gene-activities in germ-cells by means of adaptive enzymes. As he says, the model was 'intended only to suggest that it may be unsafe to consider that the occurrence of directed [non-random] mutation related to the environment can be ruled out of court *a priori*.[10]*

The isolation of the germ-cells from the rest of the body does not apply to plants, where any cell from the growing tip of the shoot can give rise to sex cells. Nor is it universal among animals—flatworms or hydra, for instance, can regenerate a whole individual, including its reproductive organs, from practically any isolated segment of their bodies. Biologists are faced with the choice of either asserting that ostriches developed callosities to sit on, just where they needed them, by pure chance—or at least to admit the theoretical possibility that some well-defined structural modifications—such as the aforementioned callosities or the thick skin on our own soles—which were acquired by generation after generation, did gradually seep through the protective filter and lead to changes in the genetic code which made them inheritable. Biochemistry does not tell us how exactly this could be achieved; but it does not exclude the possibility of a phylogenetic memory for clearly defined, vital and persistent stimulations encoded in the DNA chain as a kind of evolutionary cocktail-party effect. How else but through some process of phylogenetic memory-formation could the complex, built-in instincts of building a nest or spinning a web have arisen? Contemporary genetics has no answers to offer to the problem of the genesis of behaviour.

'It must be for experiment to decide whether such processes happen or not.' Kammerer's experiments were particularly suitable for such tests, because amphibians and reptiles, not to mention sea-siphons, are primitive creatures with great regenerative powers and genetic flexibility; and because the type of persistent stimulation to which he exposed them were realistic examples of the pressures of a changing environment which might lead to evolutionary changes. When he published his first results in his twenties, they made 'biologists all over the world sit up'. When his experiments are repeated under strict controls—and I confidently believe that this will happen sooner or later—they may have a similar, but more lasting effect.

Repetition of the *Ciona* experiments would take only a few months. *Salamandra* and *Alytes* would require at least ten years. But with automated control of temperature and humidity,

---

* Cf. also the discussion after v. Bertalanffy's and Waddington's papers in *Beyond Reductionism—The Alpbach Symposium*, ed. Koestler and Smythies (1969).

a modern research team could carry them out as a side-line with a very small expenditure of time.

Let Kammerer have the last word:

'Evolution is not just a fair dream of the last century, the century of Lamarck, Goethe and Darwin; evolution is truth—sober, delightful reality. It is not merciless selection that shapes and perfects the machinery of life; it is not the desperate struggle for survival alone which governs the world, but rather out of its own strength everything that has been created strives upwards towards light and the joy of life, burying only that which is useless in the graveyard of selection.'[11]

The Law of Seriality

Camille Flammarion, the astronomer, tells in his book *L'Inconnu et les Problèmes Psychiques* the veridical tale of Monsieur de Fontgibu and the plum pudding. A certain M. Deschamps, when a little boy in Orléans, was given by M. de Fontgibu, a visitor to his parents, a piece of plum pudding which made an unforgettable impression on him. As a young man, years later, dining in a Paris restaurant, he saw plum pudding written on the menu and promptly ordered it. But it was too late, the last portion had just been consumed by a gentleman whom the waiter discreetly pointed out—M. de Fontgibu, whom Deschamps had never seen again since that first meeting. More years passed and M. Deschamps was invited to a dinner party where the hostess had promised to prepare that rare dessert, a plum pudding. At the dinner table M. Deschamps told his little story, remarking, 'All we need now for perfect contentment is M. de Fontgibu.' At that moment the door opened and a very old, frail and distraught gentleman entered, bursting into bewildered apologies: M. de Fontgibu had been invited to another dinner party and come to the wrong address.

Flammarion belonged to that secret guild, the collectors of coincidences. Some addicts keep personal logs enriched by newspaper cuttings to prove their point that coincidences 'have a meaning'; others regard collecting as a vice in which they indulge with the guilty knowledge of sinning against the laws of rationality. Kammerer was a collector belonging to the first category; so was C. G. Jung. 'I have often come up against the phenomena in question', he wrote, 'and could convince myself how much these inner experiences meant to my patients. In most cases they were things which people do not talk about for fear of exposing themselves to thoughtless ridicule. I was amazed to see how many people have had experiences of this kind and how carefully the secret was guarded.'[1]

A typical case from Jung's own collection is the following:

A young woman I was treating had, at a critical moment, a dream in which she was given a golden scarab. While she was telling me this dream I sat with my back to the closed window. Suddenly I heard a noise behind me, like a gentle tapping. I turned round and saw a flying insect knocking against the window-pane from outside. I opened the winddow and caught the creature in the air as it flew in. It was the nearest analogy to a golden scarab that one finds in our latitudes, a scarabaeid beetle, the common rose-chafer (*Cetonia aurata*), which contrary to its usual habits had evidently felt an urge to get into a dark room at this particular moment.[2]

Kammerer started his case collection when he was twenty, and kept it up at least until *Das Gesetz der Serie* was finished in 1919. The book contains—by design or coincidence—exactly one hundred samples. Unlike most collectors with a predilection for dramatic cases, Kammerer's are nearly all drawn from trivial occurrences. The first chapter contains a motley collection of incidents from his notebooks under various headings: numbers, words, names, meeting people, letters, dreams, disasters, and so on. A few examples will illustrate his matter-of-fact, pedestrian approach:

(2a) My brother-in-law, E. von W., attended on November 4, 1910, a concert in the Bösendorf Saal (Vienna); He had seat No. 9 and his cloakroom ticket also showed No. 9.

(2b) On November 5, that is, the next day, we both attended the concert of the Philharmonic Orchestra in the Musikvereinssaal (Vienna); he had seat No. 21 (given to him by a colleague, Herr R.) and cloakroom ticket No. 21.[3]

Kammerer then comments that examples 2a and 2b have to be classified as 'a series of the second order' because the coinciding numbers of seats and cloakroom tickets recur twice on successive days; 'we shall soon see that such clusterings of series of the first order into series of the second or nth order are common, almost regular occurrences'.

(7) On September 18, 1916, my wife, while waiting for her turn in the consulting rooms of Prof. Dr. J. v. H., reads the magazine *Die Kunst*; she is impressed by some reproductions of pictures by a painter named Schwalbach, and makes a mental note to remember his name because she would like to see the originals. At that moment the door opens and the receptionist calls out to the patients: 'Is Frau Schwalbach here? She is wanted on the telephone.'[4]

(22) On July 28, 1915, I experienced the following progressive series: (a) my wife was reading about 'Mrs. Rohan', a character in the novel *Michael* by Hermann Bang; in the tramway she saw a man who looked like her friend, Prince Josef Rohan; in the evening Prince Rohan dropped in on us. (b) In the tram she overheard somebody asking the pseudo-Rohan whether he knew the village of Weissenbach on Lake Attersee, and whether it would be a pleasant place for a holiday. When she got out of the tram, she went to a delicatessen shop on the Naschmarkt, where the attendant asked her whether she happened to know Weissenbach on Lake Attersee—he had to make a delivery by mail and did not know the correct postal address.[4a]

According to popular belief, coincidences tend to come in clusters or series. Gamblers have their lucky days; at other times it is one damn thing after another. The title that Kammerer chose for his book, *Das Gesetz der Serie*, is in German almost a cliché—the equivalent of 'it never rains but it pours'. He defines his key concept as follows: 'A *Serie* manifests itself as a lawful recurrence of the same or similar things and events—a recurrence, or clustering, in time or space whereby the individual members in the sequence—as far as can be ascertained by careful analysis—are not connected by the same active cause.'[5]

The crucial phrase is '*lawful* recurrence'. Indeed the purpose of Kammerer's book was to prove that what we traditionally call a coincidence or a series of coincidences is in reality the manifestation of a universal principle in nature which operates independently from the known laws of physical causation. The 'laws of seriality' are, on this view, as fundamental as those of physical causality, but hitherto unexplored. Moreover, when Kammerer speaks of 'individual members of the sequence' he means that what we regard as isolated coincidences are merely the tips of the iceberg which happen to catch our eye, because we are conditioned, in our traditional modes of thinking, to ignore the ubiquitous manifestations of 'seriality', which otherwise would stare into our faces. In other words, if we were conscious coincidence-collectors, we would soon find ourselves transferred into a serial Wonderland universe.

Thus Kammerer set out to explore the unexplored 'laws of seriality'. It may have been an eccentric undertaking, but he went about it methodically as a zoologist devoted to taxonomy: he classified coincidences as he had classified the lizards of the Adriatic islands. The first hundred massive pages of the book are devoted to this task. If he was a Byron among toads, one might also call him the Linneaus of coincidence. In the opening chapters of the book we get a *typology* of non-causal occurrences relating to names, numbers, situations, etc., as already mentioned. This is followed by a chapter on the '*morphology* of series'. We learn to distinguish between series of the first, second, third, etc. *order*, which is determined by the number of successive 'similar or identical events': the 'Rohan' case would thus form a series of the third order (three successive recurrences). We may also distinguish series of the first, second, etc. *power*, according to the number of *parallel* concurrences. Thus the information about Kammerer's liaison with the dancer Grete Wiesenthal was contained in a letter which Lacerta wrote from Australia dated June 24, 1970; on the same day I received the same information independently from Professor Paul Weiss over dinner; half an hour later, on the same evening, the Austrian television announced that Grete Wiesenthal had died in Vienna, aged eighty-five—which makes this a 'series of the third power'. Besides 'order' and 'power', series can also be classified according to the number of their *parameters*—that is, the number of shared attributes. Thus, according to Kammerer's 'case 45', during the holiday season of 1906, Baroness Trautenburg, a spinster born in 1846, was injured by a falling tree, and at a different place, Baroness Riegershofen, a spinster born in 1846, was injured by a falling tree. Four parameters: Baroness, spinster, age, tree. A little more spectacular is Kammerer's case No. 10, concerning two young soldiers who, in 1915, were separately admitted to the military hospital of Katowitze, Bohemia. They had never met before. Both were nineteen, both had pneumonia, both were born in Silesia, both were volunteers in the Transport Corps, and both were called Franz Richter. Six parameters.

After typology and morphology we also get a *systematisation* of series: homologous and analogous series, pure and hybrid series, inverted series, alternating, cyclic, phasic series, and so on. Kammerer spent hours sitting on benches in various public parks, noting down the number of people that strolled by in both directions, classifying them by sex, age, dress, whether they carried umbrellas or parcels. He did the same on his long tram journeys from suburb to office. Then he analysed his tables and found that on every parameter they showed the typical clustering phenomena familiar to statisticians, gamblers, and insurance companies. He made, of course, the necessary allowances for such causal factors as rush-hour, weather, etc.

The theoretical value of these classificatory efforts is difficult to decide. It is easy to pick holes in the system: how many parameters has Jung's scarab knocking at the window? The quantitative assessment of similarities of form has always been a stumbling block in problems of this kind. Kammerer was not versed in the more sophisticated developments of the theory of probability. He was, therefore, unable to give a convincing answer to the classic argument of the sceptic that, given sufficient time, the most unlikely combinations are bound to turn up by pure chance—a scarab at the window, or a callosity on the ostrich. But however justified scepticism may be, this first attempt at a systematic classification of a-causal serial events may perhaps at some future date find unexpected applications. Einstein, for one, thought highly of Kammerer's book; he called it 'original and by no means absurd'.[6] Perhaps he remembered that the non-Euclidean geometries for multi-dimensional curved space, which some nineteenth-century mathematicians had invented as a perverse mathematical game, provided the basis for his cosmology.

At the end of the first, classificatory part of *Das Gesetz der Serie*, Kammerer concluded:

So far we have been concerned with the factual manifestations of recurrent series, without attempting an explanation. We have found that the recurrence of identical or similar data in contiguous areas of space or time is a simple empirical fact which has to be accepted and which cannot be explained by coincidence—or rather, which makes coincidence rule to such an extent that the concept of coincidence itself is negated.[7]

He then proceeds to the theoretical part of the book, in which he attempts to give a scientific explanation of the 'law of seriality'. The theory can be shown to be wrong in almost every important point, yet it shows tantalising flashes of intuition. It contains some astonishingly crude fallacies in physics, but leaves nevertheless a paradoxical after-taste of persuasiveness and intellectual beauty, which lingers on. Its effect is comparable to that of an Impressionist painting, which has to be viewed from a distance; if one puts one's nose into it, it dissolves into chaotic blobs.

The central idea is that, side by side with the causality of classical physics, there exists a second basic principle in the universe which tends towards unity; a force of attraction comparable to universal gravity. But while gravity acts on all mass without discrimination, this other universal force acts selectively to bring like and like together both in space and in time; it correlates by affinity, regardless whether the likeness is one of substance, form or function, or refers to symbols. The *modus operandi* of this force, the way it penetrates the trivia of everyday life, Kammerer confesses to be unable to explain[8] because it operates *ex hypothesi* outside the known laws of causality. But he points to analogies on various levels, where the same tendency towards unity, symmetry and coherence manifests itself in conventionally causal ways: from gravity and magnetism through chemical affinity, sexual attraction, biological adaptation, symbiosis, protective colouring, imitative behaviour, and so on, up to the curious observation that ageing couples, master and servant, master and dog, tend to grow more and more alike in appearance—as if they were demonstrating that they are well advanced on the road towards the 'I am thou and thou art I'.

> We thus arrive at the image of a world-mosaic or cosmic kaleidoscope, which, in spite of constant shufflings and rearrangements, also takes care of bringing like and like together.[9]

In *space* the unifying force produces clusters of events related by affinity; in *time* similarly related series; hence the rather awkward label 'seriality', as distinct from causality, which Kammerer chose for his postulated universal principle.

Series in *time*, i.e. the recurrence of similar events, he inter-
prets as manifestations of periodic or cyclic processes which
propagate themselves like waves along the time-axis in the
space-time continuum. We are, however, only aware of the
crests of the waves; these enter into consciousness and are perceived as isolated coincidences, whereas the troughs remain unnoticed (this, of course, is the exact reversal of the sceptic's
argument that out of the multitude of random events we pick
out those few which we consider significant). The waves of recurrent events may be kept in motion either by causal or by
a-causal, i.e. 'serial', forces. Examples of the former are the
planetary motions, and the periodic cycles derived from them—
seasons, tides, night and day. But the recurrent peaks and
troughs of promenaders in the park equipped with umbrellas,
and the lucky runs of the gambler, are clearly non-causally
related—they are patterns formed according to the autonomous
'laws of seriality'. Some of these are still completely obscure,
others Kammerer considers as tentatively established, devoting
a long chapter to theories about significant periods—from the
Pythagoreans' magic seven through Goethe's 'circles of good and
bad days which revolve inside me', to Swoboda's and Fliess'
twenty-three- and twenty-seven-day periods. It will be remembered that Freud, too, believed in periodicity and entertained a
protracted correspondence with Fliess on how the numbers 23
and 27 must be combined to obtain significant data for individual cycles. (Oddly enough, Kammerer mentions Freud's
name only once, in passing.)

However, Kammerer was too much of an evolutionist to
believe in Nietzsche's 'eternal return'. He realised that his
universal tendency towards repetition and symbiotic one-ness
had to be complemented by an opposite trend which would
account for the emergence of novelty and diversity. The merg-
ing of sperm and egg into a single cell is followed by the splitting
of the zygote and subsequent differentiation.

> The recurrence of a previous event [Kammerer concludes]
> is also a renewal in the literal sense in so far as it does not
> merely reproduce the past, but also carries some of the un-
> precedented with it. It is this blending of the old and new
> which conveys the experience of progression in time—
> which would be lacking if events were to return as identical
> copies of themselves, like the hands of a clock having com-

pleted their circles. Thus the progression of reality should not be compared either to circular or to pendular motion, but to motion along a three-dimensional spiral. Its turns repeat themselves and curve always in the same direction, but always at some distance along their axis: returning, yet advancing.[10]

The book ends on a quasi-Messianic note: Kammerer expressing his conviction that the study of seriality will change the destiny of man, for its action 'is ubiquitous and continuous in life, nature and cosmos. The law of seriality is the umbilical cord that connects thought, feeling, science and art with the womb of the universe which gave birth to them.'[11]

If Einstein found Kammerer's idea 'by no means absurd', it was perhaps because theoretical physicists in the age of relativity and quantum theory are accustomed to employ as a matter of routine such seemingly absurd concepts as negative mass, 'holes' in space, time flowing backwards, waves of probability and sub-atomic events to which no cause can be assigned. Another great physicist, Wolfgang Pauli—one of the greatest of our century*—went one step further. In 1950, he collaborated with C. G. Jung in developing a theory which postulated the operation of a-causal forces in nature, equal in importance to physical causality. The result was Jung's famous essay, 'Synchronicity: An A-Causal Connecting Principle', in which he quotes Kammerer at length, pays somewhat grudging tribute to him, and adopts his Law of Seriality—though he gives it a different name. Jung defines 'Synchronicity' as the 'simultaneous occurrence of two meaningfully but not causally connected events', or alternatively as a 'coincidence in time of two or more causally unrelated events which have the same or similar meaning[12] . . . equal in rank to causality as a principle of explanation'.[13] This is an almost verbatim repetition of Kammerer's definition of 'Seriality' as 'a recurrence of the same or similar things or events in time or space'—events which, as far as can be ascertained, 'are not connected by the same acting cause'. The main difference appears to be that Kammerer emphasises Seriality in time (though, of course, he includes contemporaneous coincidences in space), whereas Jung's concept of

* He postulated the so-called Pauli Exclusion Principle—one of the cornerstones of quantum theory, for which he got the Nobel Prize in 1945; he also predicted the existence of the neutrino, the strangest of all 'elementary particles', before it was discovered.

Synchronicity seems to refer only to simultaneous events—but he then explains that 'Synchronicity' is not the same as 'synchronous', but can refer to events at different times. It is psychologically interesting that Jung felt moved to coin a term and then to explain that it does not mean what it means—probably to avoid using Kammerer's term 'Seriality'.

Another difference between Kammerer's book and Jung's essay is that Jung tries to relate all a-causal phenomena to the collective unconscious and extrasensory perception, whereas Kammerer relies on analogies with physical principles such as gravity, magnetism, etc., rejecting all parapsychological explanations. Here we come to another paradox in his complex character. The most impressive and popular examples of meaningful coincidences are veridical dreams, premonitions, telepathic experiences, and so on. Kammerer believed in Seriality as an irreducible principle of life, and dismissed all parapsychological explanations as occult superstition. Nor did he apparently believe in the significance of unconscious processes, either in a Freudian or a 'serialistic' context. There are only two dreams mentioned in his collection of coincidences, both trivial and dreamt by others. The paradox is that he thought of himself as a hard-boiled philosophical materialist. He was also what one may call a devoted atheist; a freemason; a member of the Austrian Socialist Party; and a regular contributor to the *Monistische Monatshefte*, the monthly published by the German League of Monists. His last article[14] appeared in it posthumously: a description of the Darwin Museum in Moscow.

The Legend

I have mentioned in the text the famous Soviet film *Salamandra*. Unfortunately its plot is so absurd that its evidential value is zero. I saw it in Moscow in 1932 or 1933, and my memory of it is rather hazy; I shall therefore quote, as a curiosity, the only printed record of it that I could find—in an article on 'Research and Politics' by Richard Goldschmidt, published in 1949 in *Science*. That, too, was written twenty years after the event, but in this case accuracy of detail hardly matters.

Goldschmidt attended, as a guest of honour, the all-Russian Geneticist Congress held in Leningrad in 1929:

> One day, walking down the street with my friend Philip-chenko, I saw in front of a movie house a large poster of *Salamandra* decorated with pictures of this harmless animal. My surprised question was answered by my friend with an invitation to see the film. This we did, and my friend interpreted the text. . . . [The film] turned out to be nothing but a propaganda film for the doctrine of the inheritance of acquired characters. It uses the tragic figure of Kammerer, his salamanders, and mixed up with them, for the story, his midwife toads. The importance attached to the subject is revealed by the facts that none other than the then all-powerful Commissar for Education, the highly cultured and intelligent Lunacharsky, is the author of the film, that his wife plays the leading lady and that Luna-charsky, playing himself, appears in one scene. Leaving out the interwoven love story written to fit the beautiful

Mme. Lunacharsky, the plot is this: In a Central European University a young biologist (model Kammerer) is working. He is a great friend of the people and endowed with all the qualities of a Communist movie hero. Working with salamanders, he has succeeded in changing their colour by action of the environment. One day the supreme glory is achieved; the effect is inherited. The bad man of of the play, a priest, learns of this, comes to the conclusion that the discovery will spell an end to the power of the Church and the privileged classes, and decides to act. He meets at night in a church (I recognised with surprise that these pictures were taken in the glorious double cathedral of Erfurt in Thuringia) with a young prince of the blood whom he had succeeded in having appointed as assistant to Kammerer. (This is obviously a typical job for a German prince!) Here in the dark sacristy the plot is hatched. The prince (or the priest?) proposes to Kammerer that he announce his glorious discovery at a formal University meeting, and the scientist gladly accepts. During the following night the priest and the prince enter Kammerer's laboratory, to which the prince has the key, since he poses as the scientist's devoted collaborator. They open the jar in which the proof specimen of salamander is kept in alcohol, and inject the specimen with ink. [A salamander is evidently more photogenic than a toad.] Then follows the scene at the University meeting. All the professors and the president appear in academic robes, the young scientist is introduced and makes a brilliant speech announcing the final proof for the inheritance of acquired characters. When the applause has ended the priest (or was it the assistant? I am quoting entirely from memory) steps up, opens the jar, takes out the salamander, and dips it into a jar of water. All the colour runs out of the specimen. An immense uproar starts and Kammerer is ingloriously kicked out of the University as an impostor. Some time later, we see the poor young scholar walking the streets and begging with an experimental monkey which had followed him into misery. He is completely forgotten until one of his former Russian students arrives and tries to call on him. She succeeds in finding him, finally, completely down and out, in a miserable attic. She takes the train at once to Moscow and obtains an interview with Lunacharsky (this is the scene where he appears in person), who gives orders to save the victim of *bourgeois* persecution. Meanwhile, the character of Kammerer has sunk so low that he decides to make an end of it. The very moment he tries to commit suicide, the Russian student returns with Lunacharsky's message and prevents him from taking his life. The last scene shows a train in which Kammerer and the Russian saviour are riding east and a large streamer reads 'To the land of liberty'.[1]

Thus in Russia, during the years when Lysenko forced biologists to conform to his own brand of Lamarckism, Kammerer was regarded as a hero and martyr. That, of course, did not enhance his reputation in the West any more than the ballyhoo in the American Press had done. His name vanished from the textbooks. When his experiments are mentioned in some historical retrospect, they are usually dismissed with some slighting remark and with little regard for the facts. Professor C. D. Darlington, for instance, in his book *The Facts of Life* (1953), p. 223, compares the Kammerer affair with the famous case of the Tichborne heir. 'The more palpable the fraud becomes, the more devoted in their faith are the diminishing few who continue to be taken in by it: *credunt quia impossibile*. Those who read the printed evidence now may not, therefore, see that after this discussion [at the Linnean Society] Kammerer was a lost man.' In fact, as the eyewitness accounts and the correspondence in *Nature* prove (pp. 75–83), the Cambridge and Linnean meetings were a resounding success. What the passage conveys is not only that Kammerer was a 'lost man', but it also implies that the 'palpable fraud' was committed by Kammerer himself.

Any student who wishes to discover something about Kammerer will find his name mentioned in the index of Darlington's book only once, referring to the incriminating passage from which I have just quoted. He may therefore easily miss the only other mention of Kammerer in the book, not indicated in the index. It is on p. 236, consists of a single sentence, in a different context, and reads: 'The experiments of Kammerer were probably not faked by Kammerer himself.' No comment is added.

Professor H. Graham Cannon's misleading account of Kammerer's experiments I have briefly quoted before ('The *Alytes* work was first published in a short paper just before the First

World War', p. 26). Characteristic of its nasty innuendoes is a remark concerning Kammerer's motives for visiting Cambridge: 'It was just about this time that the Austrian currency collapsed and so naturally it was an invitation that Kammerer could scarcely refuse.'[2] But as the minutes of the Council of the Cambridge Natural History Society of June 7, 1923, show, Kammerer received no honorarium for his lectures. His travelling expenses from Vienna to London and back, as well as his living expenses in England from April 25 to May 11, 1923, were reimbursed, amounting to a total of £16 3s. 7d. By way of comparison: Bateson offered £50 for the safe conveyance of the specimen to London. As for historical accuracy, Cannon writes: '[The fraud] came out some months later [after the Cambridge meeting] when it was shown in America that the specimen was in fact a forgery.'[3] For 'some months' read 'three years', for 'America' read 'Vienna'. And so it goes on.

Or take this version by Richard Goldschmidt of Kammerer's death: 'Soon after this [Noble's disclosure] Kammerer . . . accepted an invitation to live in the U.S.S.R. Nothing was heard of what he did there, except that soon after he committed suicide.'[4]

What amazes the layman is that all these University professors, who only had to ask an assistant to look up the data in the back numbers of *Nature*, apparently did not feel impelled to do even that. Polemics apart, such cavalier treatment of facts would hardly be forgiven to a reporter in the popular Press.

The Unrepeatable Experiment

Bateson's principal witness against Kammerer in the case of the midwife toad was Dr. G. A. Boulenger (1858–1937), at the time Curator of Reptiles at the British Museum, Natural History. He was Belgian by birth, British by naturalisation, and was elected a Fellow of the Royal Society in 1894.

A short time before Bateson started the controversy with Kammerer by requesting the loan of an experimental *Alytes* (p. 60), he approached Boulenger and asked for his opinion on Kammerer's work. Boulenger replied:[1]

2.7.1910.
Dear Bateson,

I am of course greatly interested in the work being done in Vienna by Kammerer, and I have often wished to be able to go and see his Vivarium. I am assured that his observations are absolutely trustworthy.

In writing to me I suppose you have specially in view his results in *Vererbung erzwungener Fortpflanzungsanpassungen* (1907), which is a remarkable piece of work.

If I can be of use to you in answering any questions on which you are not clear, I shall be most pleased to do so.

Do you know what Bles [cf. Prof. Hutchinson, p. 77] is doing at Oxford? He visited the Vienna Institute some two years ago, and he told me he intended to conduct some experiments on the line of Kammerer's. I have not heard from him for a long time.

With kind regards,
Yours sincerely,
G. A. Boulenger

But under Bateson's influence, Boulenger's attitude changed. Bateson's reply is not preserved; the next item in the correspondence is dated July 11 :[2]

11.7.10.
Dear Bateson,
   I shall be very pleased to see you on Wednesday and Mrs. Boulenger hopes you will share our family dinner (we do not dress).
   That we may have plenty of time to talk over K.'s work, I shall be in at 5.30.
                                        Yours sincerely,
                                          G. A. Boulenger

The result of the family dinner was the enlistment of Boulenger and one of his sons into a thorough investigation of Kammerer's results. Three months later, Boulenger senior reports:[3]

27.10.10.
Dear Bateson,
   As a result of your enquiries respecting the markings of *Salamandra maculosa* I have put my youngest son onto a careful examination of the material in the Museum, and he has devised a method of notation of the spots and bands which he would like to submit to you. We hope to see you here when you have a moment to spare. I think the boy's results are very pertinent and answers well to my scheme of geographical distribution. But perhaps you can suggest a better graphic method.
                                        Yours sincerely,
                                          G. A. Boulenger

There are no more letters until after the war. But both Boulengers were busy attempting to refute Kammerer's experiments with salamanders and lizards, and in *Problems of Genetics* both are amply quoted by Bateson as expert witnesses for the prosecution (pp. 207–10). Thus, for instance, Kammerer had stated that in the second generation of *Salamandra maculosa* raised on yellow ground the irregular yellow spots of the animal became transformed into symmetrical longitudinal bands. Bateson denied the validity of this claim (italics his)[4]:

On returning from Vienna in 1910 I consulted Mr. G. A. Boulenger [father] in reference to the subject, and he very kindly showed me the fine series from many localities in the British Museum, and pointed out that in nature the colour-varieties can be grouped into two distinct types, one in which the yellow of the body is irregularly distributed in spots and one in which this yellow is arranged for the most part in two longitudinal bands which may be continuous or interrupted. *The spotted form is, as he showed me, an eastern variety, and the striped form belongs to western Europe.* Mr. E. G. Boulenger[5] [the son] has since published a careful account of the distribution of the two forms. The spotted he regards as the typical form, var. *typica*, and for the striped he uses the name var. *taeniata*. . . . He expresses surprise that Kammerer should not allude to these peculiarities in the geographical distribution of the two forms. He suggests further that it is more likely that some mistake occurred in Kammerer's observations than that the east European *typica* should, in the course of a generation, have been transformed into the west European *taeniata* by the influence of yellow clay soil.

This is a polite way of saying that Kammerer had substituted specimens of one variety for the other—how else could a 'mistake in observation' have occurred when the reported change took place gradually, over five years?

In the same period (1910–12) Boulenger senior was also hard at work on *Alytes*. The result was a communication to the Belgian Royal Academy, published in 1912: *Observations sur l'accouplement et la pont de l'Alyte accoucheur, 'Alytes obstetricans'*. In this paper he relates that

passing, last June, through a pretty village in the Famenne region [southern Belgium], I discovered there an abundance of *Alytes* and decided to stay there next year for a few days to observe that nocturnal Batrachian, hoping to surprise it, at long last, in the act of copulation. I carried out this plan last June, and my efforts were finally crowned with success. Accompanied by a young amateur, Mr. J. L. Monk from Birmingham, I spent seven nights in that village, and three of them were favourable to our observations. . . . The weather was, frankly, unpromising: it was very cool. Yet on the fourth evening, June the 18th, the performance we had so much wished for, was at last offered for us to view.[6]

There follow seven pages of epic description of the mating of several couples of midwife toads—observed by the light of an electric torch—and acid polemics against other naturalists (Demours, Hartmann and Lebrun) who are accused of having described it inaccurately. The attack on Kammerer is left to the end:[7]

Boulenger relates that he took away the eggs from two males immediately after they had fertilised them and attached them to themselves. He then placed the eggs into water drawn from a pond full of tadpoles. But after five or six days the eggs were dead. He concluded that the failure of these and earlier experiments 'have convinced me that the *Alytes* of France and Belgium are incapable of completing their development in water It appears that the *Alytes* of Westphalia are different, although they belong to the same species, since Kammerer, working with the latter, seems to have no difficulty in violating the laws of nature in this way.' As for Kammerer's claim

> of blackish rugosities appearing on the inner edge of the first finger—having had the opportunity to manipulate and carefully examine an *Alytes* in the act of copulation, I discovered that not one but two fingers are applied to the pubic region of the female, a fact which so far has not been stated. Since the nuptial rugosities, or mating brushes, always correspond to the manner of copulation, they ought to develop on the two internal fingers of the *Alytes* and not only on the first, as in the case of frogs whose inner finger alone is in contact with the female's breast during the embrace. I therefore believe to be entitled to cast doubt on this surprising case of atavism, which has to be added to other, even more marvellous ones in the Vienna experiments—on the reverting of *Alytes* to the aquatic mode of reproduction.[8]

Came the war and with it the breakdown of communications. Boulenger could not be accused of putting scientific objectivity above patriotic loyalty; after the war (on April 23, 1919) he wrote to Bateson:[9]

> I have not seen Kammerer's latest and have not put myself out to procure a copy as I have bound myself to ignore everything published in Germany after July 1914.

But this attitude did not prevent him from publishing in

1917, at the height of the war, a violent attack on Kammerer, whose publications he had pledged himself to ignore. It appeared in *Annals and Magazine of Natural History*, Ser. 8, Vol. XX, August 1917, under the title 'Remarks on the Midwife Toad (*Alytes obstetricans*), with reference to Dr. P. Kammerer's Publications'. It was not provoked by any new development on the *Alytes* front—the war had put an end to further experiments and Kammerer did not publish any scientific papers between 1914 and 1918. (The 'latest' paper referred to in the letter to Bateson was published in 1919.) Thus Boulenger's 1917 attack was probably motivated by other than scientific reasons. It starts:[10]

> Having recently felt bound to recommend caution in accepting the results of the experiments conducted in Vienna by Dr. Kammerer within the last fifteen years, and to express doubts as to certain alleged facts which it seems almost impossible to control . . .

The 'recently' refers to Boulenger's communication to the Belgian Academy five years earlier,[11] and no new facts are invoked in the rest of the article; but Boulenger's way of arguing is characteristic for the style of the controversy.

> From the days of Demours [1741, 1778], who first observed part of the parturition of the midwife toad, and gave a very incomplete and incorrect account of the operation, up to Kammerer's observations, only A. de l'Isle [1876] whom I have been able to confirm on all important points [1912], and Héron Royer [1886] have described this complicated and wonderful act without recourse to the imagination, which has evidently played greater or less a part in the very numerous other accounts which have appeared and to which I need not refer here. As to Kammerer himself, I feel sure he has never once carefully observed the whole operation, otherwise he would certainly have thought it worth while to allude in some way to the discrepancies between his own observations and those of de l'Isle. By not endeavouring to unravel the truth through checking the latter's account he has laid himself open to the reproach made by Spallanzani to Demours: *'Une observation si intéressante méritait bien d'être répétée, et elle me paraissait plus propre à irriter la curiosité du philosophe qu'à la satisfaire.'*[12]

Résumé: because Kammerer did not embark on polemics against de l'Isle's paper of 1876, describing the mating of *Alytes*, therefore we must conclude that Kammerer has 'never once carefully observed the whole operation'.

A further argument for doubting that Kammerer had watched *Alytes* mating is that he did not report 'on how many occasions he has spent part of the night at the *Versuchsanstalt*—a subject worth enquiring into, considering that Kammerer tells us himself (1913) that he does not reside at the *Versuchsanstalt*, but at Hütteldorf, two miles from Vienna, whilst the extent of his multifarious experiments on salamanders, *Proteus*, *Alytes*, *Hyla*, etc., would, it seems to me, have required his almost constant watch, especially after sunset, during the spring and summer. Salamanders and *Alytes* never pair in the day-time'.[13]

One is almost embarrassed to remark that although Hütteldorf (Hütteldorf-Hacking to be precise) was administratively a suburb of Vienna, it was at a travelling distance of about twenty to thirty minutes from the centre. (One might as well cast doubt on the honesty of a journalist on occasional night duty, because he lives in Hampstead.)

Boulenger also takes exception to Kammerer's statement that among the specimens sent to him by a collector in Westphalia, Dr. Hartmann, there were fourteen males and twenty-one females, although in general male *Alytes* are more numerous than females. 'I know that contradictory statements on this subject [proportion of male to female] have been made, by Leydig among others, but I can only say that my experience coincides entirely with Lataste's.'[14] Thus because Lataste wrote in 1877 that he found more males than females, while Leydig (no reference given) said the opposite, Kammerer's statement is suspect. But suspect of what? That he could not tell males from females? Irrelevancy can hardly be carried further. Yet Bateson quoted this article as 'an elaborate and destructive criticism of Dr. Kammerer's statements'.[15]

Much more serious is Boulenger's next allegation. He writes (my italics): 'A further surprising statement [of Kammerer's] in connection with the Hartmann specimens is that *all the males* should have bred three nights after their arrival (April 21). . . . Considering the protracted breeding season of the species, how can as many females as there were males have been ready to lay at the same time?' Boulenger gives two references where this alleged statement is said to occur—Kammerer's 1906 paper, p. 69; and his 1909 paper, p. 454.

Looking up the first reference, we find (my italics): 'The animals *started spawning* on April 21, when they had recovered from the journey. (Die Tiere *begannen*, nachdem sie sich vom Transport erholt hatten, am 21. April mit dem Laichen.)'

The second passage mentioned by Boulenger merely refers back to the earlier paper; but a few pages later (p. 456), talking of a different group of fifteen *Alytes* pairs, Kammerer mentions casually—what every breeder would take for granted—that they mated on various dates between April 29 and May 6.

It could hardly have been expected for readers of Boulenger's article in 1917 to look up Kammerer's original papers, and to discover that the alleged 'surprising statement' was a fabrication.

But Boulenger's main grievance was that he found himself unable to repeat Kammerer's experiment of breeding *Alytes* from eggs under water. He apparently was unable to make *Alytes* breed at all—in water or on land:

> The Zoological Society [of London] received some years ago a number of Westphalian *Alytes* purchased from the same Dr. Hartmann. Contrary to what happened when sent to Kammerer, they did not breed with us.[16]

As for Kammerer's statement that the tadpoles emerged from the eggs in water after thirteen to fifteen days: 'I could hardly, at the time I first read it, believe such a statement, having, as have others [i.e. Bateson] repeatedly tried to rear *Alytes* eggs in water, but without success. In order to satisfy myself once more, I made a further experiment in Belgium in 1912, under what I thought the best conditions, bearing in mind what Kammerer had written, taking the eggs from males immediately after they had been fertilised, and placing them in water drawn from the little pond in which they would have ultimately hatched had they been left to the care of the parent; but development stopped on the fifth or sixth day'[17]—see his report to the Belgian Academy (p. 151). Boulenger then quotes Kammerer's brief reply to that report, published in 1914: 'Young *Alytes* eggs lying in water, are in an unnatural condition which have to be compensated by artificial means—by keeping them under sterile conditions in boiled and artificially aerated water. Nevertheless some fungus spores always do get at them, and each infected egg has therefore to be carefully eliminated. Even by observing these rules of caution, more strings of eggs died and had to be thrown into the bucket than Monsieur Boulenger presumably used in his experiments, until in the end I managed to succeed with a very few eggs from a very few strings.'[18]

Boulenger quotes this reply only to accuse Kammerer that he invented this excuse when cornered:[19]

Why, it may be asked, was all this not mentioned at first, instead of letting the reader believe that the embryos underwent the whole of their development without any intervention on the part of the experimentator.

From this sample of the levity with which Kammerer relates his experiments, is it surprising if some of his statements should be challenged by those who, like myself, do not place implicit confidence in them?

But the levity—if that is the right word—is on Boulenger's side. It is simply untrue to say that Kammerer did not, from the very beginning of his experiments, explicitly point out the precariousness of *Alytes* eggs submerged in water. He did so in his first paper on the subject, published in 1906[20] (Kammerer's italics):

Before describing the next experiment, I must give some *technical indications* for the correct preservation of the eggs destined to develop without the paternal care of the male. I deposit the balls of egg-strings in glass cups on fine river sand, which has to be sterilised before use to eliminate fungus germs. According to the nature of the experiment, the sand is kept moist or wet in varying degrees. In a humid environment, the cup is closed with a well-fitting lid. If the eggs are made to develop in darkness, they are covered with sterilised blotting paper which is moistened, instead of the sand. Earth and moss, though natural media, are to be avoided, because otherwise most of the eggs are attacked by mould. Those string-balls which are to have daily baths are extracted—always at the same time of day— by means of a horn spoon from the sand bed and placed for five minutes into a cup of water. In the case of eggs reared in darkness, this manipulation is carried out in a darkroom. In spite of all these precautions, one is liable to lose a fairly

high percentage of the eggs through withering or the pro-
liferation of mould fungus so that, to be able to complete
an experiment, one has to start with substantial material.
This is particularly the case in *Experiment No. 4* to be
presently described: *The Maturation of Alytes Eggs
Deprived of Paternal Care, in Water.* . . .

And he did warn again, in his second paper on the subject,
published in 1909:[21]

It should be repeated once more that the *Alytes* eggs kept
in water are so vulnerable to begin with that on some
strings no eggs at all reach maturity, on other strings only
3 per cent to 5 per cent.

It is indeed astonishing that Boulenger has not only failed to
take these warnings into account, but denied that they had
ever been uttered. Even more astonishing: the first quote above
ends on the same page of the 1906 paper—page 69—to which
Boulenger referred as saying that all *Alytes* mated on the same
day. He thus could not have overlooked the technical instructions whose existence he denied. No wonder he failed to repeat
Kammerer's experiments; but this can hardly be construed as
an 'elaborate and destructive criticism'.

At the end of his article Boulenger repeated his criticism
about the position of the nuptial pad, quoting at length what
was said on the subject by Bateson—and yet once more Bateson's letter to Kammerer in 1910, and the latter's failure to
produce a specimen.

In spite of having 'often wished to visit the Vivarium',
Boulenger never did, and thus never had an opportunity to
acquaint himself, by first-hand experience, with Kammerer's
special techniques—or with the distance between Hütteldorf
and Vienna. After the First World War he fades out of the
picture. Kammerer's 'reply to Boulenger' (which I shall quote
in a moment) was published in his long 1919 paper which
Boulenger had 'not put himself out to procure'; he never
answered it.

His son, E. G. Boulenger, did, however, visit the Vienna
Institute in 1922. Although he too had been sceptical about
Kammerer's claims, he seems to have changed his mind and
came back with the conviction that 'if we still disbelieve, we
must assume that Przibram is a dishonest person'. He attended

Kammerer's Cambridge lecture and demonstration, but did not take part in the discussion, nor in the controversy in *Nature*.*

The last document in the Boulenger controversy is Kammerer's reply in his concluding 1919 paper on *Alytes*. It occupies forty-five pages in the *Archiv für Entwicklungsmechanik*, of which eight are devoted to replies to his critics—Boulenger, Bateson and Bauer. I shall quote the reply to Boulenger almost in full, since it sums up several vital aspects of the controversy (Kammerer's italics):

Reply to Boulenger

I have a strong aversion against polemics. That is the reason why I have not answered before various attacks on the reliability of my results and their theoretical interpretations. I postponed my reply until new results, and not just the desire to answer back, would justify it. As far as *Alytes* is concerned, this is now the case.

Boulenger is one of the few who have attempted to check my results by experiment before condemning them. He took the freshly laid and fertilised eggs from two (two!) males after observing their copulatory acts, and threw the eggs into the water of a trough ('*abreuvoir*')† in which he had seen *Alytes* tadpoles; he therefore thought the conditions ideal for his experiment.

Boulenger was lucky for apparently his eggs developed normally for as long as 5–6 days before they died. For, as I have already explained in 1914 in a short paper in reply to Boulenger, I would have expected that under these 'natural' conditions the eggs would have been totally infested with fungi. And already in 1906 I reported my own difficulties in keeping the 'water'-eggs free from saprolegniacea [a group of fungi], and recommended maintenance under as sterile conditions as possible. Boulenger passes this passage in silence. . . .

Boulenger, who by the nature of his work is understand-

---

* Boulenger, jr. (who died in 1946), seems to have been a person of great charm and humour. He travelled a lot, collecting specimens, always by sea, and mostly on foreign boats. A friend of mine, who knew him well, once asked him the reason for this preference, and got the startling reply: 'They have none of this nonsense about women and children first. . . .'

† Kammerer does not realise that '*abreuvoir*' may also mean a small pond; but the point is irrelevant.

ably unacquainted with experimental methods, did not realise what every physiologist would take for granted: that even boiled and artificially ventilated water is not immune against invasion by mould germs; each affected egg must be carefully removed, cut away from the ball (of the tangled strings) with a pair of fine scissors, and the sterilisation of the whole container must then be repeated. In spite of these precautions I had to throw away many more balls, after the embryos they contained had died, than Boulenger has ever handled. Later on, however, the results improve: in successive generations the mortality of 'water'-eggs is hardly greater than that of other *anura* who depose their eggs normally in water. But the road that leads to that stage is long. A few field observations and a basin filled with water from a puddle provide no short-cut.

Breeding *Alytes* from 'water'-eggs is the precondition for the appearance, in later generations, of nuptial pads during the mating season. And since Boulenger failed in this form of breeding, he doubts in the same breath the existence of the pads. Yet he has an added argument for his disbelief: that in 1909 I described and drew the pad only on one (the innermost) finger. Boulenger however saw that during *amplexus* [the mating embrace], in contradistinction to frogs, not one but two fingers enter into contact with the female's pubic region. This statement is derived from observing a single (!) copulating *Alytes* pair, which Boulenger picked up. He concludes: if pads did show at all on my water-mating specimens, they would have had to appear, logically on two fingers and not on one.

How can Boulenger state, with such assurance, that the position of these very timorous animals was not displaced when he took them in his hand—and without picking them up, they cannot be inspected. How can he pretend to know —assuming that his description is correct—that both fingers were used with equal strength? In view of the undoubted variability of the mating postures of *Alytes* (cf. the observations of Dähne), how can Boulenger generalise from his one single observation, and moreover extend his generalisation from land-mating to water-mating *Alytes*? I have certainly watched more *Alytes* copulations—on land and in water—than Boulenger's three cases; and I regard it as important that I watched them at close quarters in the terrarium—and not in the light of an electric torch in the gutter of a village street or in the clefts of a stone in a ruin. Yet I would not dare to make any generalised statement regarding the action of the male's fingers (which during copulation are hidden under the female); nor as to the number and position of the fingers which exert friction on the female's skin. It is even possible that the variability and extension of the nuptial pads, as apparent in the material presented in this paper, corresponds to as many modifications in the positions of the embrace.

Boulenger can, in view of the evidence, hardly doubt any longer the existence of the pads; in view of their variability he will perhaps also realise that deductions from ethological observations to morphological features require much greater caution than he has shown.

Boulenger, as already said, did not reply.

As far as the literature indicates, Boulenger and Bateson were the only zoologists who made the attempt to repeat Kammerer's experiments with *Alytes*. Bateson did not publish his results. The texts I have quoted explain why Boulenger failed; and they indicate that if Kammerer's experiments were regarded as unrepeatable and therefore suspect, the blame lies on those who tried to repeat them with inadequate methods.

The Location of the Pads

It will be remembered that after the meeting at the Linnean Society in 1923, Bateson published (*Nature*, June 2) his hitherto most violent attack on Kammerer. Among the various points which Bateson raised, the principal was that the dark mark on the skin of the critical specimen of *Alytes* was 'in the wrong place', namely, across the palm (Bateson's italics):

> I direct attention first to the fact that the structure shown did not look like a real *Brunftschwiele*. Next, I lay stress on its extraordinary position. *It was in the wrong place.* Commenting on the evidence, I pointed this out. In the embrace of *Batrachians* the palms of the hand of the male are not in contact with the female [i.e. the male turns the backs of its hands to a certain extent inwards]. To show how the hands are placed, I send a photograph (Figure 1) of a pair of *Rana agilis*, killed and preserved while coupled. Clearly the rugosities, to be effective, must be on the backs and radial sides of the digits, round the base of the thumb, as in our common frog, on the inner sides of the forearms, or in certain other positions, but not on the palms of the hands. There are, of course, minor variations, in correspondence with which the positions of the rugosities differ. But on the palm of *Alytes* they would be as unexpected as a growth of hair on the palm of a man.

To this MacBride replied on June 23. Regarding the texture of the pad, he pointed out that Bateson had omitted any reference to the microscope sections which Kammerer had shown; and that the pad, in a *Rana*, too, 'looks like a simple patch of pigment, and passing my finger over it, I could not detect the capillae by feeling'. As to the position of the pad, MacBride quoted Boulenger's description of the different positions of the pads in different species—positions including regions which never get into contact with the female.

Bateson replied very briefly on June 30, avoiding all issues except for pointing out that when Boulenger spoke of the 'inner' side of fingers showing the pad, he meant the radial, not the palmar, side.

MacBride replied on July 21 that this was quite true, 'but the callosity on the radial edge of the finger involves the palmar surface also, as Dr. Bateson may convince himself by inspecting Boulenger's figures, and as, indeed, is demonstrated to every student when he is shown the nuptial callosity of the male *Rana*'.

Bateson did not reply.

On August 18 Kammerer, back from the U.S.A., replied himself to Bateson's attack. On the crucial question he wrote:

> It is incorrect to say that the black colour is restricted to the palmar aspect. (Why should Mr. Bateson assert this when he had not seen the dorsal aspect?) Actually the pads extend to the dorsal aspect and are therefore not 'in the wrong place'.
>
> It is incorrect to say that the pad presents only 'a dark uniform surface but no capillary or thorny structures'. I send herewith an enlarged photograph in which 'rugosities' can be seen on the edge of the pad with the naked eye.

In the same issue of *Nature* (August 18), Michael Perkins also replied to Bateson. I shall quote him at some length because his letter went into more detail than any of the other correspondents', and should really have settled the issue:

> Dr. Bateson points to two details which make 'the appearance quite unlike that of any natural *Brunftschwielen*': first, that in *Alytes* there is a 'dark uniform surface . . . without the dotting or stippling so obvious in true *Brunftschwielen*'; secondly, that their position does not correspond to that of the nuptial pads in *Rana agilis*.
>
> Lataste's excellent drawings (*Ann. Sci. Nat.* (6), tom. 3, pl. II, 1876) show that a uniform darkness of the outer

layer of the pad is a characteristic feature of the *Discoglossidae* (to which *Alytes* belongs) and distinguishes them from other *Anura*. The fully developed pads of *Bufo vulgaris* are also uniformly black, and I have recently found that when such full hypertrophy of the outer epithelium is inhibited, as occasionally happens from obscure causes, it may be induced by making the male maintain a sexual embrace for a week or two. . . .

The pad of the *Alytes* 'water-breed' also resembles that of the *Discoglossid Bombinator* in having a complete layer of black pigment in the *cutis vera* which would further contribute to the uniform dark appearance which *Alytes* so well and characteristically shows. . . .

The epidermal spines are very obvious in the intact specimen, as I have repeatedly seen both with lens and binocular microscope, and as many others have witnessed in my presence. Of course, they are practically impossible to photograph on account of the glistening of a wet specimen, but a photograph at least makes clear what areas of skin are affected. These include nearly the whole of the palm, the radial surface of the inner metacarpal and part of the first phalangeal joint of the thumb, and more or less of the ventral and radial surfaces of the forearm, passing over the dorso-radial margin of the inner carpal tubercle. The *Discoglossidae* are remarkable for the very various positions in which the histological features of *Brunftschwielen* may manifest themselves, on the chin, belly, thighs, toes of the feet even; in other words, they are not necessarily dependent on contact with the female for their development. Dr. H. Gadow has shown me his sketch of the nuptial pad in *Alytes cisternasii*, Bosca., where it is developed on the tip of the thumb, extending on the palmar surface. Even in the common toad I have frequently observed the nuptial rugosity extending on to the palmar surface of the inner carpal tubercle.

Questionable as it is to draw conclusions on anatomical points by analogy from other animals, it is even more unsafe to do so as regards their habits and postures; *Alytes* does not belong even to the same suborder as *Rana agilis*.

Bateson replied to Kammerer and Perkins on September 15. He completely ignored the factual contents of Perkins' letter, which he dismissed in two sentences: 'Mr. Perkins states that "the epidermal spines are very obvious in the intact specimen". He is the only independent witness of those whose opinions have reached me, who claims to have seen anything so definite.' In his reply to Kammerer, Bateson simply repeated that the pad was 'in the wrong place on the palm'; and that it 'did not look like a nuptial pad. . . . What there may have been on the back of the hand I do not know.' This was the letter which ended with the offer of £25 for sending the *Alytes* back to London.

Przibram refused the offer in a letter to Bateson which I have quoted on page 86. I have also quoted Bateson's reply to Przibram: 'I would gladly now double my offer, etc'. But Bateson's reply to Przibram, published in *Nature* on December 22, also contained a new, and quite unfounded, accusation in the last but one sentence of the quotation which follows (Bateson's italics):

In my last letter I explained how I missed making a proper examination [of the specimen] here. Reports had varied, and I drew the inference that the nature of the black marks must be mainly a question of interpretation. Not until I saw the toad at the Linnean meeting, with the unexpected and misplaced development on the *palm of the hand*, did I discover that there was anything so positive to examine. As I thought over the incident it struck me as extraordinary that this, the real peculiarity of the specimen—which, indeed, it was set up to display—had never been mentioned by Dr. Kammerer. He left England immediately after the meeting.

The phrase '*had never been mentioned by Dr. Kammerer*' seemed to imply that Kammerer was so embarrassed by the shameful position of the pad on the palm that both in the past and during his lecture he had passed it over in silence. But Bateson's accusation happens to be demonstrably untrue. In the text of Kammerer's lecture, which he gave both in Cambridge and at the Linnean, published in *Nature* on May 12, we read (my italics):

Of the many changes which gradually appear in this water breed during the various stages of development—egg, larva, and the metamorphosed animal, young and old—I will describe only one, the above-mentioned nuptial pad of the male. At first it is confined to the innermost fingers, but in subsequent breeding seasons it extends to the other fingers, to the balls of the thumb, even to the underside of the lower arm. After spreading, it exhibits an unexpected variability, both in the same individual and between one individual and another. The variability in the same individual is shown by the characters altering from year to year and in the absence of symmetry between the right hand and the left. In one specimen the dark pad extended to all the other fingers and *almost over the whole of the left hand.*

But that is not all. By courtesy of Professor W. H. Thorpe and the Cambridge Natural History Society, I have been able to obtain the original typescript of Kammerer's lecture. There, the last sentence just quoted reads as follows (my italics): 'I have *here* a specimen in which the dark pad has extended, etc.'

(Obviously the Editor of *Nature* made one of the routine corrections in the printed version of an oral delivery.)

Bateson's repeatedly expressed surprise at the 'unexpected and misplaced development on the palm of the hand' is equally difficult to understand. A few months earlier, the younger Boulenger had, as we remember, visited the Vienna Institute and then reported, in a letter to Bateson, that he had seen the famous specimen of *Alytes* and that '*nearly the whole hand is coloured black*'. It is hard to believe that Bateson should have forgotten this. Then why pretend to be so astonished by the sight of the specimen—and why refuse to examine it, as others did?

Bateson's letter was to have a distressing influence on Przibram, but only four years later. Its immediate effect must have been to incense even that saintly man by the humiliating offer of first £25, then £50 for sending the specimen back to London, so Przibram probably did not pay much attention to the insinuations it contained. But after Kammerer's suicide Przibram went through a crisis. He had lost his oldest and most trusted collaborator, and the reputation of his Institute had been gravely damaged. For several months Przibram seems to have been unable to get over the shock. In that confused state of mind he wrote, in March, 1927, a letter to *Nature*,[1] in which he attempted to summarise the whole affair. He listed five detailed 'proofs' for the genuineness of Kammerer's results; but also seemed to imply, quoting Bateson's allegations, that concerning events from 1922 onward Kammerer was mistaken in defending the 'untoward' position of the pad on the doctored specimen:

A picture was taken in September 1922, not in the *Biologische Versuchsanstalt*, but in the photographic studio Reiffenstein, of the well-known specimen, and only from thence onwards do the mis-statements begin. On the other hand, up to 1919 the descriptions and figures of nuptial pads in *Alytes* given by Kammerer do not fit in with this specimen. . . .

We have been able to collect five proofs that in his original papers Kammerer was not hampered by the doctored specimen which has invalidated his remarks on the same subject in his books *Inheritance of Acquired Characteristics* (1924) and *Neuvererbung* (1925). The proofs are as follows:

(1) In Kammerer's original papers the nuptial pad in *Alytes* is described and pictured as being 'on the dorsal side of the thumb and on the thumb-ball' (1909, p. 516, fig. 26a), 'on the dorsal and radial side of the first fingers' (1919, p. 336), and 'across the thumb-ball on the whole internal side of the fore-arm to near the elbow' (p. 337, tb. x, fig. 2), in accord with the general appearance of nuptial pads. Even in 1923, when Kammerer showed a lantern slide of the critical specimen before the Zoological Society of London, he did not mention the disposition of the nuptial pad on the whole palm of the hand (see Bateson, *Nature*, Dec. 22, 1923, and letter to Przibram). It was not until the photographs of this specimen were used in his books (1924, p. 53, fig. 9 to the right; 1925, fig. 9, facing p. 20) that Kammerer mentions and defends the untoward position of the pad in the palm and on the outer border of the last (fourth) finger. . . .

What these rather obscure passages seem to mean is that the specimen was already doctored when, in 1922, the Reiffenstein photograph was taken, because that photograph shows the black patch on the palm, which does not fit 'the descriptions and figures of nuptial pads' as given in Kammerer's earlier papers, up to and including the last one of 1919. Up to 1919 Kammerer 'was not hampered by the doctored specimen'; whereas from 1922 onward 'the mis-statements [obviously Kammerer's mis-statements] begin'; and only after including the Reiffenstein photograph of the doctored specimen in his books (1924 and 1925) does he 'mention and defend the untoward position of the pad in the palm' which up to that date he had passed over in silence. The inference is that although the doctoring was done by somebody else, Kammerer was either unaware of it, or preferred not to mention it. (How not mentioning the conspicuous broad black mark across the palm of the specimen on display could help matters neither Przibram nor Bateson explained.)

The central issue for Przibram is the Reiffenstein photograph, and on this his letter directly contradicts what he said about it earlier on. It will be remembered that in August, 1926, he wrote a parallel report to Noble's published in *Nature*. In that report Przibram cited the photograph in question as one of the main proofs for the presence of the pads in the original state of the specimen before it had been tampered with:

> Fortunately, there are photographic plates in existence showing the state of the specimen before it left Vienna for Cambridge and during its stay in England. One of these photographs was taken in the presence of Dr. J. H. Quastel in the atelier of Reiffenstein (Vienna), and the negative travelled with Dr. Quastel to England and has been in the possession of Mr. M. Perkins (Trinity College, Cambridge) since April 1923. A reprint of it is given in Kammerer's *Neuvererbung*, Stuttgart-Heilbronn, W. Seifert-Verlag, 1925 (Abb. 9, facing p. 20).

He then goes on to quote at length testimonies from Dr. Quastel, Michael Perkins and W. Farren (a photographic expert in Cambridge) to the effect that the photograph 'shows no traces of any manipulation or retouching of the actual image of *Alytes*'. He also quotes Kammerer to the effect that 'he remembers the black substance to have been in the same place and amount, even in the living animal'. In other words, Przibram accepts that the black mark on the palm, as shown in the photograph, was genuine and that the forgery was done after the specimen's return from Cambridge.

Thus the same photograph showing the blackened palm is invoked in Przibram's 1926 letter to *Nature* as evidence for the defence, and in his 1927 letter as evidence for the prosecution. What may have caused Przibram's confusion? No new evidence had come to light since Kammerer's death. Przibram must have seen, countless times, the specimen with its blackened palm and the Reiffenstein photograph of it, yet apparently it had never before occurred to him that the pad was 'in the wrong place'. After Kammerer's suicide, however, and all the nasty rumours it provoked, his mind must have been milling round and round the events of the past until they became hopelessly muddled. He probably went back to the controversy of 1923, saw it in a different light, and only now discovered that 'the pad was in the wrong place'. No doubt he had glanced at the letters in *Nature* at the time, but he must have dismissed them as irrelevant, just as he had throughout all these years found nothing wrong with the black-handed specimen and its photograph. In this state of mind, Bateson's letter with its allegation that even in his English lectures Kammerer 'did not mention' the unspeakable location of the pad, must have seemed to Przibram to clinch the argument by making Kammerer appear to have used questionable methods in his later years. Bateson was dead, but his letter seemed to have had the effect of a minor time-bomb on Przibram's mind, getting him into a tangle of contradictions.

Even so, he never for a moment suspected that Kammerer himself committed the forgery. It is greatly to Przibram's credit that he felt no resentment toward Kammerer for the damage his suicide had caused to the Institute, and that even in his disturbed state of mind he concluded the letter with his five 'proofs' of the soundness of Kammerer's earlier results (i.e. (1) that the 1919 basic paper showed no pads in the wrong places; (2) an earlier photograph in the same paper; (3) the microtome sections showing the difference between normal and water-bred *Alytes*, and (4) showing the difference between *Alytes'* pads and pads in other specimens; (5) lastly, Kändler's sections of the rudimentary pad found in a normal specimen of *Alytes*). Przibram was fallible, but of an almost masochistic honesty; it would have been far better for the prestige of his Institute if he had simply said that an unknown person had tampered with the specimens, and left it at that.

There is one more point to be dealt with in this context. Though Kammerer did mention the blackening of 'almost the whole hand' at his English lectures, why did he not mention it in his 1919 paper, since, in his own words, the blackening was in the same place and amount in the living animal? I think the answer is implicitly contained in the following quote from the 1919 paper (Kammerer's italics):

The first specimens of *Alytes* males on heat that I found equipped with nuptial pads displayed these in the form of sharply defined, greyish-black thickenings of the upper and radial sides of the *first finger*. The second finger does not show any traces of this formation either macroscopically or under the magnifying glass. However, the more *Alytes* males, particularly those of the F4 and F5 generation, came on heat, the more frequently one could observe that the pad did not always appear in the original regions; but that, whether in the same individual or among different individuals, it has a fairly wide region at its disposal, ranging from over the thumb-ball and the whole inner side of the forearm to the proximity of the elbow, and that within this region the pads show great variability in their extension and pattern. Even asymmetries do occur: for instance, on the left only a patch of the forearm might show the pad. Figure 2, Table X shows a male of the F5 generation at the height of the mating season displaying a mighty pad over a large area of the radial side of the forearm and, incidentally, also over the thumb-balls, leaving, however, the phalanges unaffected.

Within the same individual the pad area is not irrevocably fixed to the same spot either; and we can even follow the direction of the variability: generally speaking, the area increases from one mating season to the next. If, for instance, the first pad appeared on the finger tip only, then during the second mating season the whole finger is affected, during the third the ball of the third, the fourth an adjacent area of the forearm. In other words, the variability moves in the direction of increasing areas and at the same time in the direction from distal towards proximal regions.[2]

He did not explicitly mention that in one (or several) specimens the mark had also spread across the palm; in view of the extreme variability of the pads both in *Alytes* and in other species, he probably thought this superfluous, as the passage quoted implies such a possibility. Since some toads develop pads on the tips of their fingers, others on their hind-legs, to talk of 'the wrong place' seems hardly defensible, and the whole controversy looks like making a mountain out of a mole hill. But it had to be included in this account for the sake of completeness.

Ciona

The Ciona experiments have been briefly summarised on pp. 45f.

In his 1923 Cambridge and Linnean lectures Kammerer said:

... I carried out, before 1914, what may really be an *experimentum crucis*. I have written a few words on it in my *Allgemeine Biologie*. There has been no detailed publication as yet. The subject is the Ascidian, *Ciona intestinalis*. If one cuts off the two siphons (inhalant and exhalant tubes), they grow again and become somewhat larger than they were previously. Repeated amputations on each individual specimen give finally very long tubes in which the successive new growths produce a jointed appearance of the siphons. The offspring of these individuals have also siphons longer than usual, but the jointed appearance has now been smoothed out. When the nodes are to be observed, they are due not to the operation but to interruptions in the period of growth, just as in the winter formation of rings in trees. That is to say, the particular character of the regeneration is not transferred to the progeny, but a locally increased intensity of growth is transferred. In unretouched photographs of two young *Ciona* attached by their stolons to the scratched glass of an aquarium, the upper specimen is clearly seen to be contracted; the lower is at rest and shows its monstrously long siphons in full extension. They were already there at birth, for it was bred from parents the siphons of which had become elongated by repeated amputation and growth.[1]

On November 3 *Nature* published a letter from H. Munro Fox of the Cambridge Zoological Department, of which the relevant passages read:

I repeated these amputation experiments between June and September last at the Roscoff Biological Station. The oral siphon was removed from 102 *Ciona intestinalis* which were growing attached to the walls of the tanks. The animals varied in length from 0·9 to 4·8 cm. As controls, 235 unoperated individuals were kept under observation. In none of the operated animals was there any further growth of the siphons after the original length had been reattained. . . . In 1913 it was shown at Naples that abnormally long siphons of *Ciona intestinalis* can be grown by keeping the animals in suspensions of abundant food (*Biol. Centrbl.* 1914, vol. 34, p. 429). Were this the reason for the long siphons of Dr. Kammerer's operated *Ciona*, it should have been clear from controls of unoperated animals kept in the same water.

MacBride replied to Munro Fox's letter on November 24 in *Nature* (his italics):

As Dr. Kammerer took a deep interest in the projected repetition of his experiments on *Ciona*, and wrote to me twice this summer to learn if repetition were being attempted and under what conditions, perhaps you will allow me to make some remarks on Mr. Fox's letter, as Dr. Kammerer is now in America.

Dr. Kammerer, whilst in Cambridge, wrote out a full account of the precautions to be observed in making these experiments. At that time he did not know that Mr. Fox was going to take up the work: another Cambridge biologist had undertaken to do so, but this gentleman was prevented by illness from doing the work. To him, however, Dr. Kammerer had transmitted his information. I understand—Mr. Fox will correct me if I am wrong—that Dr. Kammerer's instructions did not reach Mr. Fox. In these circumstances it is not surprising to learn that Mr. Fox failed to obtain Dr. Kammerer's results, since he has

tumbled into one of the most obvious pitfalls. It may surprise him very much to learn that *Dr. Kammerer got the
same results as he did* when, like Mr. Fox, he cut off only the
oral siphon. Since the anal siphon remains of normal
length and the reaction is of the animal as a whole, the regenerated oral siphon is of normal length also. But *when
both anal and oral siphons are amputated in a very young
animal*, then long siphons are regenerated. I have a photograph which shows an operated *Ciona* and a normal one
growing side by side in the same tank, and the contrast
between the lengths of their siphons is obvious. When Dr.
Kammerer returns from America I hope that Mr. Fox will
communicate with him and repeat the experiments, observing Dr. Kammerer's precautions, when, I feel con-
fident, he will obtain Kammerer's results.

Kammerer, on his return from America, also replied to
Munro Fox—*Nature*, December 8 (his italics):

In *Nature* of November 3, page 653, Mr. H. Munro Fox
announces that he did not succeed in repeating my results
in his *Ciona* experiments in Roscoff: amputated siphons
regained only their normal length. Mr. Fox supposes that
the extra growth in length of the siphons in my experiments
was produced by extravagant feeding, and not by the regenerative activities of the animals.

Before Mr. Fox publishes the full account of his work,
which he promises, I beg him to note the following facts,
namely:

(1) The two principal cultures (operated and control) of
my *Ciona* were placed at the same time and at the same
stage of development, with the same provision of food, in
two precisely similar aquaria, which stood beside each
other. The dimensions of these aquaria were 300 × 170 ×
100 centimetres. I did not undertake a quantitative estima-
tion of the number of micro-organisms present; but the
food available was, so far as I could see, rather on the
scanty than on the abundant side.

All the specimens in the control culture possessed short
siphons, and therefore the influence of food on the length
of siphon is excluded.

(2) I am not the first and only observer who has noted
the 'super-regeneration' of the siphons after they have been

cut off several times. Mingazzini* asserts that siphons amputated three or four times at intervals of a month became longer after each regeneration. Mingazzini was able in this way to produce artificially the local variety, 'macrosiphonica', found in the Gulf of Naples. I fully anticipated that the decisive experiment on regeneration and inheritance in *Ciona* would encounter violent contradiction. On that account I took care to construct this critical experiment out of experiments which had already been made by other investigators. That this was possible in the case of *Ciona* was one of the reasons which led me to choose this species. Indeed, I have had a predecessor (E. Schulz) also on the question of the regeneration of the 'Keimplasma' out of somatic material, though his experiments were made not on *Ciona* but on another Ascidian (*Clavellina*). The only originality which I claim is the combination of *well-known* experiments and their application to the solution of a problem of inheritance.

Barfurth,† after he had discovered (at that time in his laboratory at Dorpat) that the limbs of frog-larvae had the power of regeneration, laid stress on the superiority of one positive result as against any number of negative results. 'Even if only Dorpat tadpoles regenerated their limbs, nevertheless his result would be established.' I make the same claim for *Ciona*, 'even if only *Ciona* from Naples and Trieste grow long siphons'. Finally, have perhaps only southern populations this power?

On December 22 of the same year, Przibram wrote to Bateson refusing the '£25 offer'; in the same letter he also mentioned *Ciona*:[2]

In case you have noticed Mr. Munro Fox's letter in *Nature*, No. 2818, on *Ciona*, I would like to direct your attention to the fact that the discovery of its siphons lengthening with repeated removal was not made first by Kammerer. It was known so long ago as 1897 by Mingazzini's experiments, which were, in their turn, based on a previous observation

---

* 'Sulla regenerazione nei Tunicata', *Bolletino Soc. Nat. Napoli*, Ser. I, year 5, 1891. (An abstract of this paper appeared in the Naples *Zoologischer Jahresbericht* for 1891 under the heading 'Tunicata'.)
† 'Sind die Extremitäten der Frösche regenerationsfähig?' *Arch Entw-Mech.*, vol. I, 1894.

of our friend in common, Jacques Loeb, as he mentioned to me in 1907 during my stay in California. So I do not see how Mr. Fox's inability to reproduce the experiment allows him to deny Kammerer's success with the first generation.

On January 5, 1924, B. Stewart, a student at Trinity and member of the Cambridge Natural History Society, also wrote about *Ciona* to *Nature*. He was an amateur photographer, who had taken pictures of Kammerer's specimens and copied Kammerer's photographs of *Ciona*.

There are three photographs of *Ciona*. The first is of a single untreated specimen, the second of a group showing artificially produced *var. macrosiphonica*, and the third of two untreated offspring of the latter. In view of the various magnifications, both in the camera and from perspective, and since the whole of the animal is not visible in most cases, simple measurements would be meaningless. However, the increase of the siphon of *v. macrosiphonica* is chiefly in the direction of length, and therefore the ratios of length to breadth of the siphons provide a satisfactory method of comparing the specimens. The ratios are:
Photograph I. (Untreated, fully extended specimen.) Oral Siphon 1·9, aboral 1·65.
Photograph II. (Group.) In a single fully extended specimen, doubtless that referred to by Prof. MacBride, the ratios are 2·0 oral and 1·65 aboral. In the remainder the ratios when expanded are 4·0 to 4·3 oral and 2·0 to 4·3 aboral, and when contracted 2·4 oral and 1·9 aboral.
Photograph III. One of these two young offspring of *v. macrosiphonica* is completely expanded or nearly so, the other is quite contracted; in the former the ratios are 4·1 oral and 2·05 aboral, in the latter they are 2·35 and 1·4.
The validity of the means of comparison suggested above is shown by the ratios of length to breadth for the main part of the body lying, in all the four or five specimens in which it can be measured, between 4·1 and 4·8; i.e. the error due to varying expansion, position, and focus cannot possibly be more than 20 per cent, yet *v. macrosiphonica* shows an increase in length of the siphons of as much as 125 per cent.

On January 19 J. T. Cunningham of East London College

took issue with MacBride's explanation of the reasons for Munro Fox's negative results. He looked up Mingazzini's paper on the subject and found that 'it is distinctly stated that in some cases the buccal and cloacal siphons were cut off in different individuals, sometimes in the same individual, and that in either method a regenerated siphon showed increased length. It is to be noted that Dr. Kammerer in his letter in *Nature* of December 8 does not confirm the statement of Prof. MacBride in the issue of November 24.'

MacBride replied to this—*Nature*, February 9, 1924.

In *Nature* of January 19, p. 84, there appears a letter from Mr. Cunningham in reference to the regeneration of the siphons of *Ciona*, in which he calls in question a statement of mine in a letter in the issue of November 24. In my letter I attributed the failure of Mr. Fox to get lengthened siphons after amputation to the fact that he cut off only the oral siphon.

Mr. Cunningham says that Dr. Kammerer did not confirm my view in his subsequent letter to *Nature* (which incidentally I translated for him and sent to *Nature*). This is true; but I received afterwards a letter from Dr. Kammerer in which he explicitly agrees with my explanation and says that he had not realised that Mr. Fox had only cut off one siphon.

It appears that Mingazzini—about whose work Mr. Cunningham learnt from the letter which I translated—succeeded even when he cut off only one siphon. It may, therefore, be the case, as Dr. Kammerer suggested, that Mr. Fox failed, not because he cut off only one siphon, but because he was dealing with a northern race of *Ciona*.

The importance of the reference to Mingazzini's work lies in this, that this work unequivocally supports Dr. Kammerer's statements: many were inclined to doubt their trustworthiness after the publication of Mr. Fox's letter.

And there the controversy came to rest—as the others did. Whatever the reason for Munro Fox's failure to obtain elongated siphons, his negative result has to be weighed against the positive results obtained by Mingazzini, Jacques Loeb and shown on Kammerer's photographs. Nothing that transpired in the controversy justifies the abandonment of a line of research with far-reaching theoretical implications.

REFERENCES

CHAPTER ONE (*pages* 13 *to* 26)

1. *Neue Freie Presse*, Vienna, September 25, 1926. 2. October 30, 1926, p. 635 f. 3. K. Przibram (1959), p. 185. 4. Kammerer (1899 and 1900). 5. H. Przibram (1926), pp. 401 ff. 6. Private communication, April 6, 1970. 7. Goldschmidt (1949), p. 22. 8. The Bateson Papers. 9. MacBride (1924), p. 88. 10. Kammerer (1914b), pp. 10–11. 11. Cannon (1959), p. 45. 12. Kammerer (1906, 1909a, 1919a).

CHAPTER TWO (*pages* 27 *to* 38)

1. Goldschmidt (1949), p. 221. 2. Kammerer (1914b), pp. 15–16. 3. Hardy (1965), p. 156. 4. Waddington (1952). 5. Bergson (1911), pp. 44–5, quoted by Himmelfarb (1959), p. 369. 6. Cannon (1959), pp. ix–x. 7. Butler (1951 ed.), p. 167, quoted by Himmelfarb (1959), p. 362. 8. Butler (1879), p. 54. 9. Darlington in preface to reprint of *On the Origin of Species* (1950). 10. *Life and Letters*, II, p. 215. 11. Quoted by Himmelfarb (1959). 12. Darwin (1868). 13. Third Letter to Bentley, *Opera Omnia*, IV, p. 380. 14. Quoted by Hardy (1965), p. 157. 15. Hardy, op. cit., p. 159. 16. Darlington (1953), pp. 219–21.

CHAPTER THREE (*pages* 39 *to* 47)

1. Kammerer (1904), pp. 165–264. 2. Kammerer (1907c), pp. 99–102. 3. Kammerer (1907b), p. 34. 4. Kammerer (1923), p. 637. 5. Kammerer (1925), p. 45. 6. Ibid., p. 60. 7. Ibid., pp. 48–9. 8. Kammerer (1923a), p. 639. 9. Bateson, letter in *Nature*, May 16, 1923. 10. Bateson (1913), p. 201. 11. MacBride, letter in *Nature*, January 17, 1925. 12. Kammerer (1919a), p. 327. 13. MacBride, letter in *Nature*, December 5, 1925. 14. Goldschmidt (1940), p. 257 n. 15. Goldschmidt (1949), p. 221.

CHAPTER FOUR (*pages* 48 *to* 58)

1. The Bateson Papers, July 17, 1910. 2. H. Przibram (1926). 3. Bateson (1928). 4. Private communication, April 8, 1970. 5. Bateson (1902). 6. Hardy (1965), p. 89.

CHAPTER FIVE (*pages* 59 *to* 64)

1. Bateson (1913), p. 191. 2. Ibid., p. 190. 3. Ibid., p. 227. 4. Ibid., p. 199. 5. Ibid., p. 202. 6. The Bateson Papers.

CHAPTER SIX (*pages* 65 *to* 72)

1. Kammerer (1919a), pp. 357 ff. 2. Kammerer (1925), p. 34. 3. Darlington (1953), p. 222. 4. The Bateson Papers. 5. The Bateson Papers. 6. Ibid. 7. Ibid.

CHAPTER SEVEN (*pages* 73 *to* 86)

1. MacBride, letter in *Nature*, June 23, 1923. 2. Kammerer (1919a), p. 328. 3. Kammerer (1923a), p. 640. 4. The Bateson Papers. 5. Montagu (1970), p. 139. 6. Private communication, April 6, 1970. 7. June 23, 1923. 8. Letter in *Nature*, August 18, 1923. 9. The Bateson Papers. 10. Kammerer, letter in *Nature*, August 18, 1923. 11. Letter in *Nature*, September 15, 1923. 12. Letter in *Nature*, August 18, 1923. 13. Letter in *Nature*, December 8, 1923.

CHAPTER EIGHT (*pages* 87 *to* 97)

1. The Bateson Papers, September 20, 1923. 2. Minutes of the Council of the Cambridge Natural History Society, April 27, 1923. 3. Ibid. 4. Ibid. 5. Ibid., May 1, 1923. 6. The Bateson Papers, October 1, 1923. 7. Ibid., February 27, 1924. 8. Lecture announcement, February 16, 1924, New School for Social Research, New York. 9. Kammerer (1926b). 10. July 16, 1927.

CHAPTER NINE (*pages* 98 *to* 116)

1. Letter in *Nature*, October 16, 1926. 2. *Nature*, August 7, 1926. 3. H. Przibram, letter in *Nature*, April 30, 1927. 4. Lataste (1876). 5. The Bateson Papers, August 14, 1920. 6. April 30, 1927. 7. Letter in *Nature*, August 21, 1926. 8. April 30, 1927. 9. *Nature*, August 17, 1926. 10. H. Przibram (1926). 11. Letter in *Nature*, April 30, 1927. 12. K. Przibram (1959), p. 188. 13. Private communication, October 19, 1970. 14. Ibid., October 27, 1970. 15. *Neue Freie Presse*, September 27, 1926.

CHAPTER TEN (*pages* 117 to 122)

1. R. Wettstein, *Neue Freie Presse*, December 16, 1926. 2. Ibid. 3. Ibid. 4. September 25, 1926. 5. *Der Abend*, September 24, 1926. 6. Private communication, July 2, 1970. 7. K. Przibram, op. cit., p. 187.

EPILOGUE (*pages* 123 *to* 134)

1. Kammerer (1923a), p. 639. 2. Bateson (1924), p. 405. 3. Ibid. p. 406. 4. Bateson (1913), p. 248. 5. Johannsen (1923), p. 140. 6. June 26, 1970. 6a Thorpe (1969), p. 1. 7. von Bertalanffy (1969), pp. 66–7. 8. Koestler (1967), pp. 158–9. 9. Waddington (1957), p. 182. 10. Ibid. 11. *New York Evening Post*, February 23, 1924.

APPENDIX 1 (*pages* 135 *to* 143)

1. Jung (1960), p. 420. 2. Ibid., p. 438. 3. Kammerer (1919b), p. 24.
4. Ibid., p. 25. 4a. Ibid., p. 27. 5. Ibid., p. 36. 6. H. Przibram (1926).
7. Kammerer (1919b), p. 93. 8. Ibid., p. 137. 9. Ibid., p. 165. 10. Ibid.
p. 454. 11. Ibid., p. 456. 12. Jung, op. cit., p. 441. 13. Ibid., p. 435.
14. Kammerer (1926a).

APPENDIX 2 (*pages* 144 *to* 147)

1. Goldschmidt (1949), pp. 220–2. 2. Cannon (1959), p. 46. 3. Ibid.
4. Goldschmidt (1949), p. 221.

APPENDIX 3 (*pages* 148 *to* 159)

1. The Bateson Papers. 2. Ibid., 3. Ibid. 4. Bateson (1913), pp. 207–8.
5. E. G. Boulenger (1911), p. 323. 6. G. A. Boulenger (1912), pp.
572–3. 7. Ibid., p. 579. 8. Ibid. 9. The Bateson Papers. 10. G. A.
Boulenger (1917), pp. 173–4. 11. G. A. Boulenger (1912). 12. G. A.
Boulenger (1917), pp. 174–5. 13. Ibid., pp. 177–8. 14. Ibid., p. 176.
15. Letter in *Nature*, July 3, 1919. 16. G. A. Boulenger (1917), p. 177.
17. Ibid., p. 180 f. 18. Kammerer (1914a), p. 260. 19. G. A. Boulenger
(1917), p. 181. 20. Kammerer (1906), pp. 68–9. 21. Kammerer
(1909a), pp. 475–6.

APPENDIX 4 (*pages* 160 *to* 168)

1. April 30, 1927. 2. Kammerer (1919a), pp. 336–7.

APPENDIX 5 (*pages* 169 *to* 174)

1. Kammerer (1923a), p. 369. 2. Letter in *Nature*, December 22, 1923.

BIBLIOGRAPHY

I. WORKS BY PAUL KAMMERER*
(Not including popular articles, lectures and pamphlets on socio-cultural subjects.)

Technical Papers
Die Reptilien und Amphibien der hohen Tatra. Mitteilungen der Sektion für Naturkunde des Österreichischen Touristenklub, XI, Heft 6 und 7, 1899.
Haftzeher in Gefangenschaft. Natur and Haus, VIII, Hefte 22 und 23, 1900.
Beitrag zur Erkenntnis der Verwandtschaftsverhältnisse von Salamandra atra und maculosa. Arch. 1904. 17. 165–264.
Über die Abhängigkeit des Regenerationsvermögens der Amphibienlarven von Alter, Entwicklungsstadium und spezifischer Grösse. Arch. 1905. 19. 148–80.
Experimentelle Veränderung der Fortpflanzungstätigkeit bei Geburtshelferkröte (Alytes obstetricans) und Laubfrosch (Hyla arborea). Arch. 1906a. 22. 48–140.
Künstlicher Melanismus bei Eidechsen. Zbl. f. Physiol., Leipzig, 1906b. 20. 261–3.
Bastardierung von Flussbarsch (Perca fluviatilis L.) und Kaulbarsch (Acerina cernua L.) Arch. 1907a. 23. 511–51.
Vererbung erzwungener Fortpflanzungsanpassungen. I u. II. Mitteilung : Die Nachkommen der spätgeborenen Salamandra maculosa und der frühgeborenen Salamandra atra. Arch. 1907b. 25. 7–51.
Vererbung der erworbenen Eigenschaft habituellen Spätgebärens bei Salamandra maculosa. Zbl. f. Physiol., Leipzig, 1907c. 21. 99–102.
Erzwungene Fortpflanzungsveränderungen und deren Vererbung, etc. Zbl. f. Physiol., Leipzig, 1907d. 21. Nr. 8.
Symbiose zwischen Libellenlarve und Fadenalge. Arch. 1908a. 25. 52–81.
Regeneration sekundärer Sexualcharaktere bei den Amphibien. Arch. 1908b. 25. 82–124.

* Arch. refers to: *Archiv für Entwicklungsmechanik der Organismen,* Leipzig.

Regeneration des Dipterenflügels beim Imago. Arch. 1908c. 25. 349–60.

Vererbung erzwungener Fortpflanzungsanpassungen. III. Mitteilung: Die Nachkommen der nicht brutpflegenden Alytes obstetricans. Arch. 1909a. 28. 447–546.

Allgemeine Symbiose und Kampf ums Dasein als gleichberechtige Trieb-Kräfte der Evolution. Arch f. Rssen- u. Gesellsch.-Biol., Leipzig und Berlin, 1909b. 6. 585–608.

Vererbung erzwungener Farb- und Fortpflanzungsveränderungen. Natur, Leipzig, 1909c. 1. 94–7.

Vererbung erzwungener Farbveränderungen. I. u. II. Mitteilung: Induktion von weiblichem Dimorphismus bei Lacerta muralis, von männlichem Dimorphismus bei Lacerta fiumana. Arch. 1910a. 29. 456–98.

Die Wirkung äusserer Lebensbedingungen auf die organische Variation im Lichte der experimentellen Morphologie. Arch. 1910b. 30. I. 379–408.

Das Beibehalten jugendlich unreifer Formzustände (Neotonie und Progenese). Ergebn. d. wissensch. Med., Leipzig, 1910c. 4. 1–26.

Gregor Mendel und seine Vererbungslehre mit Rücksicht auf ihre Bedeutung für die medizinische Wissenschaft. Wien. med. Wchnschr. 1910d. 9. 2367–72.

Beweise für die Vererbung erworbener Eigenschaften durch planmässige Züchtung. Deutsche Gesellschaft für Züchtungkunde, Berlin, 1910e.

Vererbung künstlicher Zeugungs- und Farbveränderungen bei Reptilien. Vortrag Internat. Physiol. Kongress, Wien. Umschau. XV. Nr. 7. 133–56. 1911a.

Mendelsche Regeln und Vererbung erworbener Eigenschaften. Verhandl. d. Naturforsch. Ver. Brünn. 1911b. XLIX (Mendel-Festband).

Experimente über Fortpflanzung, Farbe, Augen und Körperreduktion bei Proteus anguineus Laur. (zugleich: Vererbung erzwungener Farbveränderungen. III. Mitteilung). Arch. 1911–12. 33. 349–461.

Direkt induzierte Farbanpassungen und deren Vererbung. Zeitschrift indukt. Abst.-u. Vererbungsl. 1911. IV. 279–88 und Verh. VIII Internat. Zool.-Kongress. Graz. 1912. 263–71.

Experimente über Fortpflanzung, etc. IV. Mitteilung: Das Farbkleid des Feuersalamanders, Salamandra maculosa Laurenti in seiner Abhängigkeit von der Umwelt. Arch. 1913a. 36. 4–193.

Nachweis normaler Funktion beim herangewachsenen Lichtauge des Proteus. Arch. f. d. ges. Physiol., Bonn, 1913b. 51. 1090–4.

Bemerkungen zum Laichgeschäft und der Brutpflege bei der Geburtshelferkröte (Alytes obstetricans). Blätter für Aquarien u. Terrarienkunder. 1914a. XXV. Nr. 15. 259–61.

Die Bedeutung der Vererbung erworbener Eigenschaften für Erziehung und Unterricht. Wien, 1914b.

Vererbung erzwungener Formveränderungen. I. Mitteilung: Brunst-

schwiele der Alytes-Männchen aus 'Wassereiern' (Zugleich: Verererbung erzwungener Fortpflanzungsanpassungen. V. Mitteilung). Arch. 1919a. 45. 323–70.

Die Zeichnung von Salamandra maculosa in durchfallendem farbigem Licht. Arch. 1922. 50. 79–107.

Züchtversuche über Vererbung erworbener Eigenschaften. Natur, Leipzig. 1922–3. 14. 305–11.

Breeding experiments on the inheritance of acquired characters. Nature. 1923a. 111. 637–40.

Methoden der experimentellen Variationsforschung. Hand. d. biol. Arbeitsmeth., Berlin, 1923b. Abt. 9 T.3.

Das Darwinmuseum zu Moskau. Monistische Monatshefte, October, 1926a. 11. 377–82.

Methoden und Züchtung von Reptilien und Amphibien. Pflege und Zucht weiterer wirbelloser Landtiere. In: Handb. d. biol. Arbeitsmeth., Berlin, 1928. Abt. 9 T. 1, 2, 1.

Books

Bestimmung und Vererbung des Geschlechtes bei Pflanze, Tier und Mensch. Leipzig, 1913.

Genossenschaften von Lebewesen auf Grund gegenseitiger Vorteile (Symbiose). Stuttgart, 1913c.

Allgemeine Biologie. Stuttgart, 1915 (3. Aufl. 1925).

Das Gesetz der Serie. Stuttgart, 1919b (2. Aufl. 1921).

The Inheritance of Acquired Characteristics. New York, 1924.

Neuvererbung oder Vererbung erworbener Eigenschaften. Stuttgart, 1925.

Der Artenwandel auf Inseln und seine Ursachen. Leipzig und Wien, 1926b.

2. THE KAMMERER CONTROVERSY IN *Nature**

1919   May 22: MacBride (L)
         July 3: Bateson (L)

1923   May 12: Kammerer (A)
         May 26: Cunningham (L)
         June 2: Bateson (L)
         June 23: MacBride (L)
         June 30: Bateson (L)
         July 21: MacBride (L)
         July 28: Cunningham (L)
         August 18: Kammerer (L)
                      Perkins (L)
         September 8: MacBride (L)
                      Sir Arthur Keith (L)
         September 15: Bateson (L)
         November 3: Munro Fox (L)

* A refers to articles, L to letters.

November 24: MacBride (L)
December 8: Kammerer (L)
December 15: Cunningham (L)
December 22: Przibram (L)
1924 January 5: Stewart (L)
January 19: Cunningham (L)
February 9. MacBride (L)
May 17: Dover (L)
June 5: Calman (L)
July 26: Cunningham (L)
September 6: Przibram (L)
September 27: Calman (L)
1925 January 10: MacBride (L)
November 28: MacBride (L)
December 5: MacBride (L)
1926 January 9: Cunningham (L)
February 13: MacBride (L)
March 6: MacBride (L)
March 20: Cunningham (L)
May 29: Cunningham (L)
August 7: Noble (A)
Przibram (A)
August 21: MacBride (L)
October 2: Obituary notice
October 9: Noble (L)
October 16: Obituary by Przibram
October 30: Obituary (unsigned)
November 6: MacBride (L)
1927 April 30: Przibram (L)
Kiplinger (L)
May 14: MacBride (L)
July 16: MacBride (A)

3. OTHER WORKS MENTIONED IN THIS BOOK

Bateson, W., *Mendel's Principles of Heredity : A Defence*. Cambridge, 1902.
——*Problems of Genetics*. New Haven and London, 1913.
——*Naturalist*. Cambridge, 1928.
——*Letters from the Steppe*. London, 1928.
The Bateson Papers, Library of the American Philosophical Society, Philadelphia.
Bergson, H., *Creative Evolution*. Tr. A. Mitchell. London, 1911.
von Bertalanffy, L., in *Beyond Reductionism*. See Koestler and Smythies, ed., 1969.
Boulenger, E. G., *Proc. Zool. Soc.*, 1911, p. 323.

Boulenger, G. A., 'Observations sur l'accouplement et la ponte de l'Alyte accoucheur, *Alytes obstetricans*'. Academie Royale de Belgique, *Bulletin de la Classe des Sciences*, 1912. Nos. 9–10. 570–9.

——'Remarks on the Midwife Toad (*Alytes obstetricans*), with reference to Dr. P. Kammerer's Publications.' *Annals and Magazine of Natural History*, August, 1917. Ser. 8, Vol. XX. 173–84.

Butler, Samuel, *Evolution Old and New*. 1879.

——*Notebooks*. Ed. G. Keynes and B. Hill. New York, 1951.

Cannon, H. Graham, *The Evolution of Living Things*. Manchester, 1958.

——*Lamarck and Modern Genetics*. Manchester, 1959.

Darlington, C. D., in preface to *On the Origin of Species*. Reprint of 1st ed., London, 1950.

——*The Facts of Life*. London, 1953.

Darwin, Charles, *The Variation of Animals and Plants under Domestication*. 2 vols. London, 1868.

——*On the Origin of Species*. Reprint of 1st ed., London, 1950.

Dobzhansky, T., *The Biology of Ultimate Concern*. New York, 1967.

Flammarion, C., *L'Inconnu et les Problemes Psychiques*. Paris, 1900.

Focke, W., *Die Pflanzen Mischlinge*, 1881.

Goldschmidt, R., *The Material Basis of Evolution*. New Haven, 1940.

——'Research and Politics'. *Science,* March 4, 1949.

Hardy, Sir A., *The Living Stream*. London, 1965.

Himmelfarb, G., *Darwin and the Darwinian Revolution*. London, 1959.

Johanssen, W., 'Some Remarks about Units in Heredity'. *Hereditas*, 1923. IV, p. 140.

Jung, C. G., *The Structure and Dynamics of the Psyche*. The Collected Works, Vol. 8. Tr. R. F. C. Hull. London, 1960.

Kammerer, P.—see Bibliography Part, 1.

Koestler, A., *The Sleepwalkers*. London and New York, 1959.

——*The Act of Creation*. London and New York, 1964.

——*The Ghost in the Machine*. London and New York, 1967.

——ed. with J. R. Smythies, *Beyond Reductionism—New Perspectives in the Life Sciences. The Alpbach Symposium*. London and New York, 1969.

Lamarck, J. P., *Philosophie Zoologique*, 2 vols., ed. C. Martins. 2nd ed., Paris, 1873.

Lataste, F., *Ann. Sci. Nat.*, 1876. (6), tom. 3.

MacBride, E. W., *An Introduction to Heredity*. London, 1924.

MacDougall, W., *British Journal of Psychology*. 1927. Vol. 17. 268–304.

Mahler-Werfel, A., *Mein Leben*. Frankfurt a. M., 1960.

Mendel, G., *Experiments in Plant Hybridisation*, 1865.

Montagu, I., *The Youngest Son*. London, 1970.

Newton, Sir I., *Opera Omnia*. London, 1779–85.

Noble, G. K., *The Biology of Amphibia*. New York and London, 1931.

Przibram, H., 'Paul Kammerer als Biologe'. *Monistische Monatshefte*, November, 1926, pp. 401–5.

——*Experimental-Zoologie*. 7 *Bände*. Vienna and Leipzig: Deuticke, 1907–30.

Przibram, K., 'Hans Przibram' in *Grosse Osterreicher, Band XIII*. Zürich, Leipzig, Vienna, 1959.

St. Hilaire, G., *Philosophie Anatomique*. Paris, 1818.

Salisbury, F. B., 'Natural Selection and the Complexity of the Gene'. *Nature*, October 25, 1969, pp. 342–3.

Semon, R., *The Mneme*. London, 1921.

Smith, J. M., *Nature*, 1970. 225. 563.

Smythies, J. R.—see Koestler (1969).

Sonneborn, T. M., 'Gene Action in Development'. *Proc. Royal Soc. of London B*, December, 1970. Vol. 176. No. 1044.

Spetner, L. M., 'Natural Selection versus Gene Uniqueness'. *Nature*, June 6, 1970, pp. 948–9.

Thorpe, W. H., in *Beyond Reductionism*, see Koestler and Smythies ed.

Waddington, C. H., *The Listener*, London, February 13, 1952.

——*The Strategy of the Genes*. London, 1957.

Whyte, L. L. *Internal Factors in Evolution*. London, 1966.

INDEX